Geometry of Submanifolds and Homogeneous Spaces

Geometry of Submanifolds and Homogeneous Spaces

Special Issue Editors

Andreas Arvanitoyeorgos
George Kaimakamis

MDPI • Basel • Beijing • Wuhan • Barcelona • Belgrade

MDPI

Special Issue Editors

Andreas Arvanitoyeorgos
University of Patras
Greece

George Kaimakamis
Hellenic Army Academy
Greece

Editorial Office
MDPI
St. Alban-Anlage 66
4052 Basel, Switzerland

This is a reprint of articles from the Special Issue published online in the open access journal *Symmetry* (ISSN 2073-8994) in 2019 (available at: https://www.mdpi.com/journal/symmetry/special_issues/ Geometry_Submanifolds_Homogeneous_Spaces).

For citation purposes, cite each article independently as indicated on the article page online and as indicated below:

LastName, A.A.; LastName, B.B.; LastName, C.C. Article Title. *Journal Name* **Year**, *Article Number*, Page Range.

ISBN 978-3-03928-000-1 (Pbk)
ISBN 978-3-03928-001-8 (PDF)

Contents

About the Special Issue Editors

Andreas Arvanitoyeorgos is a Professor of Mathematics in the Department of Mathematics at the University of Patras, Greece. He obtained his Ph.D. from the University of Rochester, USA. His research interests are in differential geometry.

George Kaimakamis is Professor of Mathematics in the Hellenic Army Academy. He obtained his Ph.D. from the University of Patras, Greece. His research interests are in differential geometry, operation research, and mathematical economics.

Preface to "Geometry of Submanifolds and Homogeneous Spaces"

The present Special Issue of Symmetry is devoted to two important areas of global Riemannian geometry, namely submanifold theory and the geometry of Lie groups and homogeneous spaces. Submanifold theory originated from the classical geometry of curves and surfaces. Homogeneous spaces are manifolds that admit a transitive Lie group action, historically related to F. Klein's Erlangen Program and S. Lie's idea to use continuous symmetries in studying differential equations.

In this Special Issue, we provide a collection of papers that not only reflect some of the latest advancements in both areas, but also highlight relations between them and the use of common techniques. Applications to other areas of mathematics are also considered.

Andreas Arvanitoyeorgos, George Kaimakamis
Special Issue Editors

Article

Chen Inequalities for Warped Product Pointwise Bi-Slant Submanifolds of Complex Space Forms and Its Applications

Akram Ali * and Ali H. Alkhaldi

Department of Mathematics, College of Science, King Khalid University, 9004 Abha, Saudi Arabia; ahalkhaldi@kku.edu.sa
* Correspondence: akramali133@gmail.com; Tel.: +966-554-146-618

Received: 24 December 2018; Accepted: 31 January 2019; Published: 11 February 2019

Abstract: In this paper, by using new-concept pointwise bi-slant immersions, we derive a fundamental inequality theorem for the squared norm of the mean curvature via isometric warped-product pointwise bi-slant immersions into complex space forms, involving the constant holomorphic sectional curvature c, the Laplacian of the well-defined warping function, the squared norm of the warping function, and pointwise slant functions. Some applications are also given.

Keywords: mean curvature; warped products; compact Riemannian manifolds; pointwise bi-slant immersions; inequalities

---

1. Introduction

In the submanifolds theory, creating a relationship between extrinsic and intrinsic invariants is considered to be one of the most basic problems. Most of these relations play a notable role in submanifolds geometry. The role of immersibility and non-immersibility in studying the submanifolds geometry of a Riemannian manifold was affected by the pioneering work of the Nash embedding theorem [1], where every Riemannian manifold realizes an isometric immersion into a Euclidean space of sufficiently high codimension. This becomes a very useful object for the submanifolds theory, and was taken up by several authors (for instance, see [2–15]). Its main purpose was considered to be how Riemannian manifolds could always be treated as Riemannian submanifolds of Euclidean spaces. Inspired by this fact, Nolker [16] classified the isometric immersions of a warped product decomposition of standard spaces. Motivated by these approaches, Chen started one of his programs of research in order to study the impressibility and non-immersibility of Riemannian warped products into Riemannian manifolds, especially in Riemannian space forms (see [11,17–19]). Recently, a lot of solutions have been provided to his problems by many geometers (see [18] and references therein).

The field of study which includes the inequalities for warped products in contact metric manifolds and the Hermitian manifold is gaining importance. In particular, in [17], Chen observed the strong isometrically immersed relationship between the warping function f of a warped product $M_1 \times_f M_2$ and the norm of the mean curvature, which isometrically immersed into a real space form.

Theorem 1. *Let $\widetilde{M}(c)$ be a m-dimensional real space form and let $\varphi : M = M_1 \times_f M_2$ be an isometric immersion of an n-dimensional warped product into $\widetilde{M}(c)$. Then:*

$$\frac{\Delta f}{f} \leq \frac{n^2}{4n_2}||H||^2 + n_1 c, \tag{1}$$

where $n_i = dimM_i$, $i = 1, 2$, and Δ is the Laplacian operator of M_1 and H is the mean curvature vector of M^n. Moreover, the equality holds in (1) if, and only if, φ is mixed and totally geodesic and $n_1 H_1 = n_2 H_2$ such that H_1 and H_2 are partially mean curvatures of M_1 and M_2, respectively.

In [2,5,20–31], the authors discuss the study of Einstein, contact metrics, and warped product manifolds for the above-mentioned problems. Furthermore, in regard to the collections of such inequalities, we referred to [12] and references therein. The motivation came from the study of Chen and Uddin [32], which proved the non-triviality of warped-product pointwise bi-slant submanifolds of a Kaehler manifold with supporting examples. If the sectional curvature is constant with a Kaehler metric, then it is called complex space forms. In this paper, we consider the warped-product pointwise bi-slant submanifolds which isometrically immerse into a complex space form, where we then obtain a relationship between the squared norm of the mean curvature, constant sectional curvature, the warping function, and pointwise bi-slant functions. We will announce the main result of this paper in the following.

Theorem 2. *Let $\widetilde{M}^{2m}(c)$ be the complex space form and let $\varphi : M^n = M_1^{n_1} \times_f M_2^{n_2} \to \widetilde{M}^{2m}(c)$ be an isometric immersion from warped product pointwise bi-slant submanifolds into $\widetilde{M}^{2m}(c)$. Then, the following inequality is satisfied:*

$$\Delta(lnf) \leq ||\nabla \ln f||^2 + \frac{n^2}{4n_2}||H||^2 + \frac{n_1 c}{4} - \frac{3c}{4n_2}\left(n_1 \cos^2 \theta_1 + n_2 \cos^2 \theta_2\right), \tag{2}$$

where θ_1 and θ_2 are pointwise slant functions along M_1 and M_2, respectively. Furthermore, ∇ and Δ are the gradient and the Laplacian operator on $M_1^{n_1}$, respectively, and H is the mean curvature vector of M^n. The equality case holds in (2) if and only if φ is a mixed totally geodesic isometric immersion and the following satisfies

$$\frac{H_1}{H_2} = \frac{n_2}{n_1}$$

where H_1 and H_2 are the mean curvature vectors along $M_1^{n_1}$ and $M_2^{n_2}$, respectively.

As an application of Theorem 2 in a compact orientated Riemannian manifold with a free boundary condition, we prove that:

Theorem 3. *Let $M^n = M_1^{n_1} \times_f M_2^{n_2}$ be a compact, orientate warped product pointwise bi-slant submanifold in a complex space form $\widetilde{M}^{2m}(c)$ such that $M_1^{n_1}$ is a n_1-dimensional and $M_2^{n_2}$ is a n_2-dimensional pointwise slant submanifold $\widetilde{M}^{2m}(c)$. Then, M^n is simply a Riemannian product if, and only if:*

$$||H||^2 \geq \frac{c}{n^2}\left(3n_1 \cos^2 \theta_1 + 3n_2 \cos^2 \theta_2 - n_1 n_2\right), \tag{3}$$

where H is the mean curvature vector of M^n. Moreover, θ_1 and θ_2 are pointwise slant functions.

By using classifications of pointwise bi-slant submanifolds which were defined in [32], we derived similar inequalities for warped product pointwise pseudo-slant submanifolds [33], warped product pointwise semi-slant submanifolds [34], and CR-warped product submanifolds [17] in a complex space form as well.

2. Preliminaries and Notations

An almost complex structure J and a Riemannian metric g, such that $J^2 = -I$ and $g(JX, JY) = g(X, Y)$, for $X, Y \in \mathfrak{X}(\widetilde{M})$, where I denotes the identity map and $\mathfrak{X}(\widetilde{M})$ is the space containing vector fields tangent to $\widetilde{M}$, then (M, J, g) is an almost Hermitian manifold. If the almost complex structure satisfied $(\widetilde{\nabla}_U J)V = 0$, for any $U, V \in \mathfrak{X}(\widetilde{M})$ and $\widetilde{\nabla}$ is a Levi-Cevita connection $\widetilde{M}$. In this case, $\widetilde{M}$ is called the Kaehler manifold. A complex space form of constant holomorphic sectional curvature c is denoted by $\widetilde{M}^{2m}(c)$, and its curvature tensor $\widetilde{R}$ can be expressed as:

$$\widetilde{R}(U, V, Z, W) = \frac{c}{4}\bigg(g(U, Z)g(V, W) - g(V, Z)g(U, W) + g(U, JZ)g(JV, W)$$

$$- g(V, JZ)g(U, JW) + 2g(U, JV)g(JZ, W) \bigg), \tag{4}$$

for every $U, V, Z, W \in \mathfrak{X}(\widetilde{M}^{2m}(c))$. A Riemannian manifold $\widetilde{M}^m$ and its submanifold M, the Gauss and Weingarten formulas are defined by $\widetilde{\nabla}_U V = \nabla_U V + h(U, V)$, and $\widetilde{\nabla}_U \xi = -A_\xi U + \nabla_U^\perp \xi$, respectively for each $U, V \in \mathfrak{X}(M)$ and for the normal vector field ξ of M, where h and A_ξ are denoted as the second fundamental form and shape operator. They are related as $g(h(U, V), N) = g(A_N U, V)$. Now, for any $U \in \mathfrak{X}(M)$ and for the normal vector field ξ of M, we have:

$$(i) \ JU = PU + FU, \quad (ii) \ J\xi = t\xi + f\xi, \tag{5}$$

where $PU(t\xi)$ and $FU(f\xi)$ are tangential to M and normal to M, respectively. Similarly, the equations of Gauss are given by:

$$R(U, V, Z, W) = \widetilde{R}(U, V, Z, W) + g(h(U, W), h(V, Z)) - g(h(U, Z), h(V, W)). \tag{6}$$

for all U, V, Z, W are tangent M, where R and $\widetilde{R}$ are defined as the curvature tensor of $\widetilde{M}^m$ and M^n, respectively.

The mean curvature H of Riemannian submanifold M^n is given by

$$H = \frac{1}{n} trace(h).$$

A submanifold M^n of Riemannian manifold $\widetilde{M}^m$ is said to be totally umbilical and totally geodesic if $h(U, V) = g(U, V)H$ and $h(U, V) = 0$, for any $U, V \in \mathfrak{X}(M)$, respectively, where H is the mean curvature vector of M^n. Furthermore, if $H = 0$, them M^n is minimal in $\widetilde{M}^m$.

A new class called a "pointwise slant submanifold" has been studied in almost Hermitian manifolds by Chen-Gray [35]. They provided the following definitions of these submanifolds:

Definition 1. *[35] A submanifold M^n of an almost Hermitian manifold $\widetilde{M}^{2m}$ is a pointwise slant if, for any non-zero vector $X \in \mathfrak{X}(T_x M)$ and each given point $x \in M^n$, the angle $\theta(X)$ between JX and tangent space $T_x M$ is free from the choice of the nonzero vector X. In this case, the Wirtinger angle become a real-valued function and it is non-constant along M^n, which is defined on $T^* M$ such that $\theta : T^* M \to \mathbb{R}$.*

Chen-Gray in [35] derived a characterization for the pointwise slant submanifold, where M^n is a pointwise slant submanifold if, and only if, there exists a constant $\lambda \in [0, 1]$ such that $P^2 = -\cos^2 \theta I$, where P is a (1,1) tensor field and I is an identity map. For more classifications, we referred to [35].

Following the above concept, a pointwise bi-slant immersion was defined by Chen-Uddin in [18], where they defined it as follows:

Definition 2. *A submanifold M^n of an almost Hermitian manifold $\widetilde{M}^{2m}$ is said to be a pointwise bi-slant submanifold if there exists a pair of orthogonal distributions $\mathcal{D}_{\theta_1}$ and $\mathcal{D}_{\theta_2}$, such that:*

(i) $TM^n = \mathcal{D}_{\theta_1} \oplus \mathcal{D}_{\theta_2}$;
(ii) $J\mathcal{D}_{\theta_1} \perp \mathcal{D}_{\theta_2}$ and $J\mathcal{D}_{\theta_2} \perp \mathcal{D}_{\theta_1}$;

(iii) Each distribution $\mathcal{D}_{\theta_i}$ is a pointwise slant with a slant function $\theta_i : T^*M \to \mathbb{R}$ for $i = 1, 2$.

Remark 1. *A pointwise bi-slant submanifold is a bi-slant submanifold if each slant functions $\theta_i : T^*M \to \mathbb{R}$ for $i = 1, 2$. are constant along M^n (see [13]).*

Remark 2. *If $\theta_1 = \frac{\pi}{2}$ or $\theta_2 = \frac{\pi}{2}$, then M^n is called a pointwise pseudo-slant submanifold (see [33]).*

Remark 3. *If $\theta_1 = 0$ or $\theta_2 = 0$, in this case, M^n is a coinciding pointwise semi-slant submanifold (see [14,34]).*

Remark 4. *If $\theta_2 = \frac{\pi}{2}$ and $\theta_1 = 0$, then M^n is CR-submanifold of the almost Hermitian manifold.*

In this context, we shall define another important Riemannian intrinsic invariant called the scalar curvature of $\widetilde{M}^m$, and denoted at $\widetilde{\tau}(T_x\widetilde{M}^m)$, which, at some x in $\widetilde{M}^m$, is given:

$$\widetilde{\tau}(T_x\widetilde{M}^m) = \sum_{1 \leq \alpha < \beta \leq m} \widetilde{K}_{\alpha\beta}, \tag{7}$$

where $\widetilde{K}_{\alpha\beta} = \widetilde{K}(e_\alpha \wedge e_\beta)$. It is clear that the first equality (7) is congruent to the following equation, which will be frequently used in subsequent proof:

$$2\widetilde{\tau}(T_x\widetilde{M}^m) = \sum_{1 \leq \alpha < \beta \leq m} \widetilde{K}_{\alpha\beta}, \ 1 \leq \alpha, \beta \leq n. \tag{8}$$

Similarly, scalar curvature $\widetilde{\tau}(L_x)$ of L-plan is given by:

$$\widetilde{\tau}(L_x) = \sum_{1 \leq \alpha < \beta \leq m} \widetilde{K}_{\alpha\beta}, \tag{9}$$

An orthonormal basis of the tangent space T_xM is $\{e_1, \cdots e_n\}$ such that $e_r = (e_{n+1}, \cdots e_m)$ belong to the normal space $T^\perp M$. Then, we have:

$$h_{\alpha\beta}^r = g(h(e_\alpha, e_\beta), e_r),$$

$$||h||^2 = \sum_{\alpha, \beta = 1}^{n} g(h(e_\alpha, e_\beta), h(e_\alpha, e_\beta)). \tag{10}$$

Let $K_{\alpha\beta}$ and $\widetilde{K}_{\alpha\beta}$ be the sectional curvatures of the plane section spanned by e_α and e_β at x in a submanifold M^n and a Riemannian manifold $\widetilde{M}^m$, respectively. Thus, $K_{\alpha\beta}$ and $\widetilde{K}_{\alpha\beta}$ are the intrinsic and extrinsic sectional curvatures of the span $\{e_\alpha, e_\beta\}$ at x. Thus, from the Gauss Equation (6)(i), we have:

$$2\tau(T_xM^n) = K_{\alpha\beta} = 2\widetilde{\tau}(T_xM^n) + \sum_{r=n+1}^{m} \left(h_{\alpha\alpha}^r h_{\beta\beta}^r - (h_{\alpha\beta}^r)^2 \right)$$

$$= \widetilde{K}_{\alpha\beta} + \sum_{r=n+1}^{m} \left(h_{\alpha\alpha}^r h_{\beta\beta}^r - (h_{\alpha\beta}^r)^2 \right). \tag{11}$$

The following consequences come from (6) and (11), as:

$$\tau(T_xM_1^{n_1}) = \sum_{r=n+1}^{m} \sum_{1 \leq i < j \leq n_1} \left(h_{ii}^r h_{jj}^r - (h_{ij}^r)^2 \right) + \widetilde{\tau}(T_xM_1^{n_1}). \tag{12}$$

Similarly, we have:

$$\tau(T_x M_2^{n_2}) = \sum_{r=n+1}^{m} \sum_{n_1+1 \leq a < b \leq n} \left(h_{aa}^r h_{bb}^r - (h_{ab}^r)^2 \right) + \widetilde{\tau}(T_x M_2^{n_2}). \tag{13}$$

Assume that $M_1^{n_1}$ and $M_2^{n_2}$ are two Riemannian manifolds with their Riemannian metrics g_1 and g_2, respectively. Let f be a smooth function defined on $M_1^{n_1}$. Then, the warped product manifold $M^n = M_1^{n_1} \times_f M_2^{n_2}$ is the manifold $M_1^{n_1} \times M_2^{n_2}$ furnished by the Riemannian metric $g = g_1 + f^2 g_2$, which defined in [36]. When considering that the $M^n = M_1^{n_1} \times_f M_2^{n_2}$ is the warped product manifold, then for any $X \in \mathfrak{X}(M_1)$ and $Z \in \mathfrak{X}(M_2)$, we find that:

$$\nabla_Z X = \nabla_X Z = (X \ln f)Z. \tag{14}$$

Let $\{e_1, \cdots e_n\}$ be an orthonormal frame for M^n; then, summing up the vector fields such that:

$$\sum_{i=1}^{n_1} \sum_{j=1}^{n_2} K(e_\alpha \wedge e_\beta) = \sum_{\alpha=1}^{n_1} \sum_{\beta=1}^{n_2} \left((\nabla_{e_\alpha} e_\alpha) \ln f - e_\alpha (e_\beta \ln f) - (e_\alpha \ln f)^2 \right).$$

From (Equation (3.3) in [11]), the above equation implies that:

$$\sum_{\alpha=1}^{n_1} \sum_{\beta=1}^{n_2} K(e_\alpha \wedge e_\beta) = n_2 \left(\Delta(\ln f) - ||\nabla(\ln f)||^2 \right) = \frac{n_2 \Delta f}{f}. \tag{15}$$

Remark 5. *A warped product manifold $M^n = M_1^{n_1} \times_f M_2^{n_2}$ is said to be trivial or a simple Riemannian product manifold if the warping function f is constant.*

3. Main Inequality for Warped Product Pointwise Bi-Slant Submanifolds

To obtain similar inequalities like Theorem 1, for warped product pointwise bi-slant submanifolds of complex space forms, we need to recall the following lemma.

Lemma 1. *[10] Let $a_1, a_2, \ldots a_n, a_{n+1}$ be $n+1$ be real numbers with*

$$(\sum_{i=1}^{n} a_i)^2 = (n-1)(\sum_{i=1}^{n} a_i^2 + a_{n+1}), n \geq 2.$$

Then $2a_1.a_2 \geq a_3$ holds if and only if $a_1 + a_2 = a_3 = \cdots = a_k$.

Proof of Theorem 2. If substitute $X = Z = e_\alpha$ and $Y = W = e_\beta$ for $1 \leq \alpha, \beta \leq n$ in (4), and (6), taking summing up then

$$\sum_{\alpha,\beta=1}^{n} \widetilde{R}(e_\alpha, e_\beta, e_\alpha, e_\beta) = \frac{c}{4} \left(n(n-1) + 3 \sum_{\alpha,\beta=1}^{n} g^2(Je_\alpha, e_\beta) \right). \tag{16}$$

As M^n is a pointwise bi-slant submanifold, we defined an adapted orthonormal frame as $n = 2d_1 + 2d_2$ follows $\{e_1, e_2 = \sec\theta_1 P e_1, \ldots, e_{2d_1-1}, e_{2d_1} = \sec\theta_1 P e_{2d_1-1}, \ldots, e_{2d_1+1}, e_{2d_1+2} = \sec\theta_2 P e_{2d_1+1}, \ldots, e_{2d_1+2d_2-1}, e_{2d_1+2d_2} = \sec\theta_2 P e_{2d_1+2d_2-1}\}$. Thus, we defined it such that $g(e_1, Je_2) = -g(Je_1, e_2) = g(Je_1, \sec\theta_1 P e_1)$, which implies that $g(e_1, Je_2) = -\sec\theta_1 g(Pe_1, Pe_1)$. $\square$

Following ((2.8) in [32]), we get $g(e_1, Je_2) = \cos\theta_1 g(e_1, e_2)$. Therefore, we easily obtained the following relation:

$$g^2(e_\alpha, Je_\beta) = \begin{cases} \cos^2\theta_1, & for\ each\ \alpha = 1, \ldots, 2d_1 - 1, \\ \cos^2\theta_2, & for\ each\ \beta = 2d_1 + 1, \ldots, 2d_1 + 2d_1 - 1. \end{cases}$$

Hence, we have:

$$\sum_{\alpha,\beta=1}^{n} g^2(Je_\alpha, e_\beta) = (n_1 \cos^2\theta + n_2 \cos^2\theta). \tag{17}$$

Following from (17), (16), and (6), we find that:

$$2\tau = \frac{c}{4} n(n-1) + \frac{c}{4}\left(3n_1 \cos^2\theta_1 + 3n_2 \cos^2\theta_2\right) + n^2 ||H||^2 - ||h||^2. \tag{18}$$

Let us assume that:

$$\delta = 2\tau - \frac{c}{4} n(n-1) - \frac{c}{4}\left(3n_1 \cos^2\theta_1 + 3n_2 \cos^2\theta_2\right) - \frac{n^2}{2} ||H||^2. \tag{19}$$

Then, from (19), and (18), we get:

$$n^2 ||H||^2 = 2(\delta + ||h||^2). \tag{20}$$

Thus, from an orthogonal frame $\{e_1, e_2, \cdots e_n\}$, the proceeding equation takes the new form:

$$\left(\sum_{r=n+1}^{2m} \sum_{i=1}^{n} h_{AA}^r\right)^2 = 2\left(\delta + \sum_{r=n+1}^{2m} \sum_{i=1}^{n} (h_{AA}^r)^2 + \sum_{r=n+1}^{2m+1} \sum_{i<j=1}^{n} (h_{AB}^r)^2 \right.$$
$$\left. + \sum_{r=n+1}^{2m} \sum_{A,B=1}^{n} (h_{AB}^r)^2\right). \tag{21}$$

This can be expressed in more detail, such as:

$$\frac{1}{2}\left(h_{11}^{n+1} + \sum_{A=2}^{n_1} h_{AA}^{n+1} + \sum_{l=n_1+1}^{n} h_{ll}^{n+1}\right)^2 = \delta + (h_{11}^{n+1})^2 + \sum_{A=2}^{n_1} (h_{AA}^{n+1})^2 + \sum_{l=n_1+1}^{n} (h_{ll}^{n+1})^2$$
$$- \sum_{2 \leq B \neq q \leq n_1} h_{BB}^{n+1} h_{qq}^{n+1} - \sum_{n_1+1 \leq l \neq s \leq n} h_{ll}^{n+1} h_{ss}^{n+1}$$
$$+ \sum_{A<B=1}^{n} (h_{AB}^{n+1})^2 + \sum_{r=n+1}^{2m} \sum_{A,B=1}^{n} (h_{AB}^r)^2. \tag{22}$$

Assume that $a_1 = h_{11}^{n+1}$, $a_2 = \sum_{A=2}^{n_1} h_{AA}^{n+1}$, and $a_3 = \sum_{l=n_1+1}^{n} h_{ll}^{n+1}$. Then, applying Lemma 1 in (22), we derive:

$$\frac{\delta}{2} + \sum_{A<B=1}^{n} (h_{AB}^{n+1})^2 + \frac{1}{2}\sum_{r=n+1}^{2m} \sum_{A,B=1}^{n} (h_{AB}^r)^2 \leq \sum_{2 \leq B \neq q \leq n_1} h_{BB}^{n+1} h_{qq}^{n+1}$$
$$+ \sum_{n_1+1 \leq l \neq s \leq n} h_{ll}^{n+1} h_{ss}^{n+1}. \tag{23}$$

with equality holds in (23) if and only if

$$\sum_{A=2}^{n_1} h_{AA}^{n+1} = \sum_{B=n_1+1}^{n} h_{BB}^{n+1}. \tag{24}$$

On the other hand, from (15), we have:

$$\frac{n_2 \Delta f}{f} = \tau - \sum_{1 \le A < B \le n_1} K(e_A \wedge e_B) - \sum_{n_1+1 \le l < q \le n} K(e_l \wedge e_q). \tag{25}$$

Then from (6) and the scalar curvature for the complex space form (11), we get:

$$n_2 \frac{\Delta f}{f} = \tau - \frac{n_1(n_1-1)c}{8} - \frac{3n_1 c}{4} \cos^2 \theta_1 - \sum_{r=n+1}^{2m} \sum_{1 \le A \ne B \le n_1} (h_{AA}^r h_{BB}^r - (h_{AB}^r)^2)$$
$$- \frac{n_2(n_2-1)c}{8} - \frac{3n_2 c}{4} \cos^2 \theta_2 - \sum_{r=n+1}^{2m} \sum_{n_1+1 \le l \ne q \le n} (h_{ll}^r h_{qq}^r - (h_{lq}^r)^2). \tag{26}$$

Now from (23) and (26), we have:

$$n_2 \frac{\Delta f}{f} \le \rho - \frac{n(n-1)c}{8} + \frac{n_1 n_2 c}{4} - \frac{3n_1 c}{4} \cos^2 \theta_1 - \frac{\delta}{2} - \frac{3n_2 c}{4} \cos^2 \theta_2. \tag{27}$$

Using (19) in the above equation and relation $\frac{\Delta f}{f} = \Delta(\ln f) - ||\nabla \ln f||^2$, we derive:

$$n_2 \left(\Delta(\ln f) - ||\nabla \ln f||^2 \right) \le \frac{n^2}{4} ||H||^2 + \frac{c}{4} \left(n_1 n_2 + 3n_1 \cos^2 \theta_1 + 3n_2 \cos^2 \theta_2 \right). \tag{28}$$

which implies inequality. The equality sign holds in (2) if, and only if, the leaving terms in (23) and (24) imply that:

$$\sum_{r=n+2}^{2m} \sum_{B=1}^{n_1} h_{BB}^r = \sum_{r=n+2}^{2m} \sum_{A=n_1+1}^{n_1} h_{AA}^r = 0, \tag{29}$$

and $n_1 H_1 = n_2 H_2$, where H_1 and H_2 are partially mean curvature vectors on $M_1^{n_1}$ and $M_2^{n_2}$, respectively. Moreover, also from (23), we find that

$$h_{AB}^r = 0, \quad \text{for each } 1 \le A \le n_1$$
$$n_1 + 1 \le B \le n$$
$$n + 1 \le r \le 2m. \tag{30}$$

This shows that φ is a mixed, totally geodesic immersion. The converse part of (30) is true in a warped product pointwise bi-slant into the complex space form. Thus, we reached our promised result.

Consequences of Theorem 2

Inspired by the research in [6,34] and using the Remark 3 in Theorem 2 for pointwise semi-slant warped product submanifolds, we obtained:

Corollary 1. *Let $\varphi : M^n = M_1^{n_1} \times_f M_2^{n_2} \to \widetilde{M}^{2m}(c)$ be an isometric immersion from the warped product pointwise semi-slant submanifold into a complex space form $\widetilde{M}^{2m}(c)$, where $M_1^{n_1}$ is the holomorphic and $M_2^{n_2}$ is the pointwise slant submanifolds of $\widetilde{M}^{2m}(c)$. Then, we have the following inequality:*

$$\Delta(lnf) \leq ||\nabla \ln f||^2 + \frac{n^2}{4n_2}||H||^2 + \frac{n_1 c}{4} - \frac{3c}{4n_2}\left(n_1 + n_2 \cos^2 \theta\right), \tag{31}$$

where $n_i = dim M_i$, $i = 1, 2$. Furthermore, ∇ and Δ are the gradient and the Laplacian operator on $M_1^{n_1}$, respectively, and H is the mean curvature vector of M^n. The equality sign holds in (31) if, and only if, $n_1 H_1 = n_2 H_2$, where H_1 and H_2 are the mean curvature vectors along $M_1^{n_1}$ and $M_2^{n_2}$, respectively, and φ is a mixed, totally geodesic immersion.

From the motivation studied in [14,34], we present the following consequence of Theorem 2 by using the Remark 2 for a nontrivial warped product pointwise pseudo-slant submanifold of a complex space, such that:

Corollary 2. *Let $\varphi : M^n = M_1^{n_1} \times_f M_2^{n_2} \to \widetilde{M}^{2m}(c)$ be an isometric immersion from a warped product pointwise pseudo-slant submanifold into a complex space form $\widetilde{M}^{2m}(c)$, such that $M_1^{n_1}$ is a totally real and $M_2^{n_2}$ is a pointwise slant submanifold of $\widetilde{M}^{2m}(c)$. Then, we have the following inequality:*

$$\Delta(lnf) \leq ||\nabla \ln f||^2 + \frac{n^2}{4n_2}||H||^2 + \frac{n_1 c}{4} - \frac{3c}{4}\cos^2 \theta, \tag{32}$$

where $n_i = dim M_i$, $i = 1, 2$. Furthermore, ∇ and Δ are the gradient and the Laplacian operator on $M_1^{n_1}$, respectively, and H is the mean curvature vector of M^n. The equality condition holds in (32) if, and only if, the following satisfies

$$\frac{H_1}{H_2} = \frac{n_2}{n_1}$$

: where H_1 and H_2 are the mean curvature vectors along $M_1^{n_1}$ and $M_2^{n_2}$, respectively, and φ is a mixed, totally geodesic isometric immersion.

Corollary 3. *Let $\varphi : M^n = M_1^{n_1} \times_f M_2^{n_2} \to \widetilde{M}^{2m}(c)$ be an isometric immersion from a warped product pointwise pseudo-slant submanifold into a complex space form $\widetilde{M}^{2m}(c)$, such that $M_1^{n_1}$ is a pointwise slant and $M_2^{n_2}$ is a totally real submanifold of $\widetilde{M}^{2m}(c)$. Then, we have the following:*

$$\Delta(lnf) \leq ||\nabla \ln f||^2 + \frac{n^2}{4n_2}||H||^2 + \frac{n_1 c}{4} - \frac{3n_1 c}{4n_2}\cos^2 \theta, \tag{33}$$

where $n_i = dim M_i$, $i = 1, 2$. Furthermore, ∇ and Δ are the gradient and the Laplacian operator on $M_1^{n_1}$, respectively, and H is the mean curvature vector of M^n. This equally holds in (33) if, and only if, φ is a mixed, totally geodesic isometric immersion and the following satisfies

$$\frac{H_1}{H_2} = \frac{n_2}{n_1},$$

, where H_1 and H_2 are the mean curvature vectors along $M_1^{n_1}$ and $M_2^{n_2}$, respectively.

Similarly, using Remark 4 and from [17], we got the following result from Theorem 2:

Corollary 4. *Let $\varphi : M^n = M_1^{n_1} \times_f M_2^{n_2} \to \widetilde{M}^{2m}(c)$ be an isometric immersion from a CR-warped product into a complex space form $\widetilde{M}^{2m}(c)$, such that $M_1^{n_1}$ is a holomorphic submanifold and $M_2^{n_2}$ is a totally real submanifold of $\widetilde{M}^{2m}(c)$. Then, we get the following:*

$$\Delta(lnf) \leq ||\nabla \ln f||^2 + \frac{n^2}{4n_2}||H||^2 + \frac{n_1c}{4} - \frac{3n_1c}{4n_2}, \tag{34}$$

where $n_i = dimM_i$, $i = 1, 2$. Furthermore, ∇ and Δ are the gradient and the Laplacian operator on $M_1^{n_1}$, respectively, and H is the mean curvature vector of M^n. The same holds in (34) if, and only if, φ is mixed and totally geodesic, and $n_1 H_1 = n_2 H_2$, where H_1 and H_2 are the mean curvature vectors on $M_1^{n_1}$ and $M_2^{n_2}$, respectively.

In particular, if both pointwise slant functions $\theta_1 = \theta_2 = \frac{\pi}{2}$, then M^n is becomes a totally real warped product submanifold—thus, we obtain:

Corollary 5. *Let $\varphi : M^n = M_1^{n_1} \times_f M_2^{n_2} \to \widetilde{M}^{2m}(c)$ be an isometric immersion from an n-dimensional, totally real warped product submanifold into a 2m-dimensional complex space form $\widetilde{M}^{2m}(c)$, where $M_1^{n_1}$ and $M_2^{n_2}$ are totally real submanifolds of $\widetilde{M}^{2m}(c)$. Then, we have the following:*

$$\Delta(lnf) \leq ||\nabla \ln f||^2 + \frac{n^2}{4n_2}||H||^2 + \frac{n_1c}{4}, \tag{35}$$

where $n_i = dimM_i$, $i = 1, 2$ and Δ is the Laplacian operator on $M_1^{n_1}$. The same holds in (35) if, and only if, φ is mixed and totally geodesic, and the following satisfies

$$\frac{H_1}{H_2} = \frac{n_2}{n_1},$$

where H_1 and H_2 are the mean curvature vectors on $M_1^{n_1}$ and $M_2^{n_2}$, respectively.

Proof of Theorem 3. In this direction, we consider the warped product pointwise bi-slant submanifolds as a compact oriented Riemannian manifold without boundary. If the inequality (2) holds:

$$\Delta(lnf) - ||\nabla \ln f||^2 \leq \frac{n^2}{4n_2}||H||^2 + \frac{c}{4n_2}\left(n_1n_2 - 3n_1\cos^2\theta_1 - 3n_2\cos^2\theta_2\right). \tag{36}$$

Since M^n is a compact oriented Riemannian submanifold without boundary, then we have following formula with respect to the volume element:

$$\int_{M^n} \Delta f dV = 0. \tag{37}$$

From the hypothesis of the theorem, M^n is a compact warped product submanifold; then from (37), we derive:

$$\int_M \left(\frac{c}{4n_2}\left(3n_1\cos^2\theta_1 + 3n_2\cos^2\theta_2 - n_1n_2\right) - \frac{1}{4n_2}\sum_{i=1}^n (h_{ii}^{n+1})^2\right)dV \leq \int_M (||\nabla \ln f||^2)dV. \tag{38}$$

Now, we assume that M^n is a Riemannian product, and the warping function f must be constant on M^n. Then, from (38), we get the inequality (3). $\square$

Conversely, let the inequality (3) hold; then from (38), we derive:

$$0 \leq \int_{M^n} (||\nabla \ln f||^2) \leq 0.$$

The above condition implies that $||\nabla \ln f||^2 = 0$, where this means that f is a constant function on M^n. Hence, M^n is simply a Riemannian product of $M_1^{n_1}$ and $M_2^{n_2}$, respectively. Thus, the theorem is proved. We give some other important corollaries as consequences of Theorem 2, as follows:

Corollary 6. *Let* $M^n = M_1^{n_1} \times_f M_2^{n_2}$ *be a warped product pointwise bi-slant submanifold of a complex space form* $\widetilde{M}^{2m}(c)$ *with warping function* f, *such that* $n_1 = dimM_1$ *and* $n_2 = dimM_2$. *If* φ *is an isometrically minimal immersion from warped product* M^n *into* $\widetilde{M}^{2m}(c)$, *then we obtain:*

$$\Delta(lnf) \leq ||\nabla \ln f||^2 + \frac{c}{4n_2}\left(n_1 n_2 - 3n_1 \cos^2\theta_1 - 3n_2 \cos^2\theta_2\right). \tag{39}$$

Corollary 7. *Let* $M^n = M_1^{n_1} \times_f M_2^{n_2}$ *be a warped product pointwise bi-slant submanifold of a complex space form* $\widetilde{M}^{2m}(c)$ *with warping function* f, *such that* $n_1 = dimM_1$ *and* $n_2 = dimM_\theta$. *Then, there is no existing minimal isometric immersion* φ *from warped product* M^n *into* $\widetilde{M}^{2m}(c)$ *with:*

$$\Delta(lnf) > ||\nabla \ln f||^2 + \frac{c}{4n_2}\left(n_1 n_2 - 3n_1 \cos^2\theta_1 - 3n_2 \cos^2\theta_2\right). \tag{40}$$

Author Contributions: All authors made an equal contribution to draft the manuscript.

Funding: The authors extended their appreciation to the Deanship of Scientific Research at King Khalid University, for funding this work through research groups program under grant number R.G.P.1/79/40.

Acknowledgments: The authors would like to thank the referees for their stick criticism and suggestions on this paper to improve the quality. The authors are grateful to Siraj Uddin for his useful comments, discussions and constant encouragement which improved the paper.

Conflicts of Interest: The authors declare no conflict of interest.

References

1. Nash, J. The imbedding problem for Riemannian manifolds. *Ann. Math.* **1956**, *63*, 20–63. [CrossRef]
2. Ali, A.; Lee; J.W.; Alkhaldi, A.H. Geometric classification of warped product submanifolds of nearly Kaehler manifolds with a slant fiber. *Int. J. Geom. Methods. Mod. Phys.* **2018**. [CrossRef]
3. Ali, A.; Laurian-Ioan, P. Geometric classification of warped products isometrically immersed in Sasakian space forms. *Math. Nachr.* **2018**, *292*, 234–251.
4. Ali, A.; Laurian-Ioan, P. Geometry of warped product immersions of Kenmotsu space forms and its applications to slant immersions. *J. Geom. Phys.* **2017**, *114*, 276–290. [CrossRef]
5. Ali, A.; Ozel, C. Geometry of warped product pointwise semi-slant submanifolds of cosymplectic manifolds and its applications. *Int. J. Geom. Methods Mod. Phys.* **2017**, *14*, 175002. [CrossRef]
6. Ali, A.; Uddin, S.; Othmam, W.A.M. Geometry of warped product pointwise semi-slant submanifold in Kaehler manifolds. *Filomat* **2017**, *31*, 3771–3788. [CrossRef]
7. Chen, B.-Y. A general inequality for submanifolds in complex-space-forms and its applications. *Arch. Math.* **1996**, *67*, 519–528. [CrossRef]
8. Chen, B.-Y. Mean curvature and shape operator of isometric im-mersions in real-space-forms. *Glasgow Math. J.* **1996**, *38*, 87–97. [CrossRef]
9. Chen, B.-Y. Relations between Ricci curvature and shape operatorfor submanifolds with arbitrary codimension. *Glasgow Math. J.* **1999**, *41*, 33–41. [CrossRef]
10. Chen, B.-Y.; Dillen, F.; Verstraelen, L.; Vrancken, L. Characterization of Riemannian space forms, Einstein spaces and conformally flate spaces. *Proc. Am. Math. Soc.* **1999**, *128*, 589–598.

11. Chen, B.-Y. On isometric minimal immersions from warped products into real space forms. *Proc. Edinb. Math. Soc.* **2002**, *45*, 579–587. [CrossRef]
12. Chen, B.-Y. *Pseudo-Riemannian Geometry, δ-Invariants and Applications*; World Scientific: Hackensack, NJ, USA, 2011.
13. Uddin, S.; Chen, B.-Y.; Al-Solamy, F.R. Warped product bi-slant immersions in Kaehler manifolds. *Mediterr. J. Math.* **2017**, *14*, 95. [CrossRef]
14. Uddin, S.; Stankovic, M.S. Warped product submanifolds of Kaehler manifolds with pointwise slant fiber. *Filomat* **2018**, *32*, 35–44. [CrossRef]
15. Uddin, S.; Al-Solamy, F.R.; Shahid, M.H.; Saloom, A. B.-Y. Chen's inequality for bi-warped products and its applications in Kenmotsu manifolds. *Mediterr. J. Math.* **2018**, *15*, 193. [CrossRef]
16. Nolker, S. Isometric immersions of warped products. *Differ. Geom. Appl.* **1996**, *6*, 1–30. [CrossRef]
17. Chen, B.-Y. Geometry of warped product CR-submanifolds in Kaehler manifolds. *Monatsh. Math.* **2001**, *133*, 177–195. [CrossRef]
18. Chen, B.-Y. *Differential Geometry of Warped Product Manifolds and Submanifolds*; World Scientific: Hackensack, NJ, USA, 2017.
19. Uddin, S.; Al-Solamy, F.R. Warped product pseudo-slant immersions in Sasakian manifolds. *Publ. Math. Debrecen* **2017**, *91*, 331–348. [CrossRef]
20. Al-Solamy, F.R.; Khan, V.A.; Uddin, S. Geometry of warped product semi-slant submanifolds of nearly Kaehler manifolds. *Results Math.* **2017**, *71*, 783–799. [CrossRef]
21. Alqahtani, L.S.; Uddin, S. Warped product pointwise pseudo-plant submanifolds of locally product Riemannian manifolds. *Filomat* **2018**, *32*, 423–438. [CrossRef]
22. Chen, B.-Y. A general optimal inequality for warped products in complex projective spaces and its applications. *Proc. Jpn. Acad. Ser. A* **2003**, *79*, 89–94. [CrossRef]
23. Defever, F.; Mihai, I.; Verstraelen, L. B. Y. Chen's inequality for C-totally real submanifolds in Sasakian space forms. *Boll. Unione Matematica Ital. B* **1997**, *11*, 365–374.
24. Decu, S.; Haesen, S.; Verstraelen, L.; Vîlcu, G.E. Curvature invariants of Statistical Submanifolds in Kenmotsu Statistical manifolds of constant φ-sectional curvature. *Entropy* **2018**, *20*, 529. [CrossRef]
25. He, G.; Liu, H.; Zhang, L.Optimal inequalities for the casorati curvatures of submanifolds in generalized space forms endowed with semi-symmetric non-metric connections. *Symmetry* **2016**, *8*, 113. [CrossRef]
26. Liaqat, M.; Laurian, P.; Othman, W.A.M.; Ali, A.; Gani, A.; Ozel, C. Estimation of inequalities for warped product semi-slant submanifolds of Kenmotsu space forms. *J. Inequal. Appl.* **2016**, *2016*, 239. [CrossRef]
27. Li, J.; He, G.; Zhao, P. On Submanifolds in a Riemannian Manifold with a Semi-Symmetric Non-Metric Connection. *Symmetry* **2017**, *9*, 112. [CrossRef]
28. Matsumoto, K.; Mihai, I. Warped product submanifolds in Sasakian space forms. *SUT J. Math.* **2002**, *38*, 135–144.
29. Uddin, S.; Chi, A.Y.M. Warped product pseudo-slant submanifolds of nearly Kaehler manifolds. *An. Stiintifice Univ. Ovidius Constanta* **2011**, *19*, 195–204.
30. Uddin, S.; Al-Solamy, F.R.; Khan, K.A. Geometry of warped product pseudo-slant submanifolds in nearly Kaehler manifolds. *An. Ştiinţ. Univ. Al. I. Cuza Iaşi. Mat* **2016**, *3*, 223–234.
31. Zhang, P.; Zhang, L. Casorati inequalities for submanifolds in a Riemannian manifold of quasi-constant curvature with a semi-symmetric metric connection. *Symmetry* **2016**, *8*, 19. [CrossRef]
32. Chen, B.-Y.; Uddin, S. Warped product pointwise bi-slant submanifolds of Kaehler manifolds. *Publ. Math. Debrecen* **2018**, *92*, 183–199. [CrossRef]
33. Srivastava, S.K.; Sharma, A. Pointwise pseudo-slant warped product submanifolds in a Kaehler Manifold. *Mediterr. J. Math.* **2017**, *14*, 20. [CrossRef]
34. Sahin, B. Warped product pointwise semi-slant submanifolds of Kaehler manifolds. *Port. Math.* **2013**, *70*, 252–268. [CrossRef]
35. Chen, B.-Y.; Gray, O.J. Pointwise slant submanifolds in almost Hermitian manifolds. *Turk. J. Math.* **2012**, *79*, 630–640.
36. Bishop, R.L.; Neil, B.O. Manifolds of negative curvature. *Trans. Am. Math. Soc.* **1969**, *145*, 1–9. [CrossRef]

symmetry

MDPI

Article

On a New type of Tensor on Real Hypersurfaces in Non-Flat Complex Space Forms

George Kaimakamis [†] and Konstantina Panagiotidou [*,†]

Faculty of Mathematics and Engineering Sciences, Hellenic Army Academy, Vari, 16673 Attiki, Greece; gmiamis@gmail.com
* Correspondence: konpanagiotidou@gmail.com
† These authors contributed equally to this work.

Received: 17 March 2019; Accepted: 16 April 2019; Published: 18 April 2019

Abstract: In this paper the notion of *-Weyl curvature tensor on real hypersurfaces in non-flat complex space forms is introduced. It is related to the *-Ricci tensor of a real hypersurface. The aim of this paper is to provide two classification theorems concerning real hypersurfaces in non-flat complex space forms in terms of *-Weyl curvature tensor. More precisely, Hopf hypersurfaces of dimension greater or equal to three in non-flat complex space forms with vanishing *-Weyl curvature tensor are classified. Next, all three dimensional real hypersurfaces in non-flat complex space forms, whose *-Weyl curvature tensor vanishes identically are classified. The used methods are based on tools from differential geometry and solving systems of differential equations.

Keywords: real hypersurfaces; non-flat complex space forms; *-Ricci tensor; *-Weyl curvature tensor

---

1. Introduction

A Kahler manifold $\tilde{N}$ is a complex manifold of complex dimension n and real dimension 2n, which is equipped with

- a complex structure J defined $J : T\tilde{N} \to \tilde{N}$, where $T\tilde{N}$ is the tangent space of $\tilde{N}$, satisfying relations $J^2 = -Id$ and $\tilde{\nabla} J = 0$, i.e., J is parallel with respect to the Levi-Civita connection $\tilde{\nabla}$ of $\tilde{N}$
- and a Riemanian metric G that is compatible with J, i.e., $G(JX, JY) = G(X, Y)$ for all tangent X, Y on $\tilde{N}$.

The pair (J, G) is called *Kahler structure*. A Kahler manifold of constant holomorphic sectional curvature c is called *complex space form*. Complete and simply connected complex space forms depending on the value of holomorphic sectional curvature c are analytically isometric to complex projective space $\mathbb{C}P^n$ if $c > 0$, to complex hyperbolic space $\mathbb{C}H^n$ if $c < 0$ or to complex Euclidean space $\mathbb{C}^n$ if $c = 0$. This paper focuses on complex space forms with $c \neq 0$ denoted by $M_n(c)$ and called *non-flat complex space forms*. Furthermore, $c = 4$ in the case of $\mathbb{C}P^n$ and $c = -4$ in the case of $\mathbb{C}H^n$.

A submanifold M in a non-flat complex space form $M_n(c)$ of real codimension equal to 1 is called *real hypersurface*. Let N be a locally defined unit normal vector on M. The Kahler structure (J, G) of the ambient space $M_n(c)$ induces on M an *almost contact metric structure* (ϕ, ξ, η, g) defined in the following way

- $\xi = -JN$ is the *structure vector field*,
- ϕ is a skew-symmetric tensor field of type (1,1) called *structure tensor field* and defined to be the tangential component of $JX = \phi X + \eta(X)N$, for all tangent vectors X to M,
- η is a 1-form and is given by the relation $\eta(X) = g(X, \xi)$ for all tangent vectors X to M,
- g is the induced Riemannian metric on M.

A big class of real hypersurfaces in $M_n(c)$ are *Hopf hypersurfaces*, which are real hypersurfaces whose structure vector field ξ is an eigenvector of the shape operator A of M, i.e.,

$$A\xi = \alpha\xi, \tag{1}$$

where $\alpha = g(A\xi, \xi)$ and is called *Hopf principal curvature*.

Takagi classified homogeneous real hypersurfaces in complex projective space $\mathbb{C}P^n$, $n \geq 2$. The real hypersurfaces are divided into six types:

- type (A) which are either geodesic hyperspheres of radius r, $0 < r < \frac{\pi}{2}$, or tubes of radius r, with $0 < r < \frac{\pi}{2}$ over totally geodesic $\mathbb{C}P^k$, $1 \leq k \leq n - 2$,
- type (B) which are tubes of radius r, $0 < r < \frac{\pi}{4}$, over the complex quadric Q^{n-1},
- type (C) which are tubes over the Serge embedding of $\mathbb{C}P^1 \times \mathbb{C}P^m$, with $2m + 1 = n$ and $n \geq 5$,
- type (D) which are tubes over the Plücker embedding of the Grassmann manifold $G_{2,5}$ and $n = 9$,
- type (E) which are tubes over the canonical embedding of the Hermitian symmetric space $SO(10)/U(5)$ and $n = 15$, where $SO(n)$ is a subgroup of $O(n)$ of dimension n, which consists of all the orthogonal matrices with determinant equal 1. (see [1–3]).

The above real hypersurfaces are Hopf ones with constant principal curvatures (see [4]).

In the case of the ambient space being the complex hyperbolic $\mathbb{C}H^n$, Montiel in [5] studied real hypersurfaces with two constant principal curvatures. Additionally, he proved that such real hypersurfaces are Hopf ones. Berndt in [6] classified Hopf hypersurfaces with constant principal curvatures in $\mathbb{C}H^n$, $n \geq 2$. The following list includes the Hopf hypersurfaces with constant principal curvatures.

- type (A) which are either horospheres, or geodesic hyperspheres, or tubes over totally geodesic complex hyperbolic hyperplane, or tubes over totally geodesic $\mathbb{C}H^k$, $1 \leq k \leq n - 2$,
- type (B) which are tubes over totally geodesic real hyperbolic space $\mathbb{R}H^2$ (type (B)).

All of them are homogeneous ones, but in contrast to the case of complex projective space, it is proved that there are also non-Hopf hypersurfaces in $\mathbb{C}H^n$ which are homogeneous.

Let $\tilde{M}$ be a Riemannian manifold of dimension m and g its Riemannian metric. Then the *Weyl curvature tensor* $W(X, Y)Z$ of $\tilde{M}$ is given by

$$
\begin{aligned}
W(X,Y)Z &= R(X,Y)Z + \frac{1}{m-2}[g(SX,Z)Y - g(SY,Z)X + g(X,Z)SY - g(Y,Z)SX] \\
&\quad - \frac{\rho}{(m-1)(m-2)}[g(X,Z)Y - g(Y,Z)X], \quad \text{for all } X, Y, Z \text{ tangent to } M,
\end{aligned}
$$

with R being the Riemannian curvature tensor, S being the Ricci tensor and ρ being the scalar curvature of $\tilde{M}$. If $m = 3$ then $W(X,Y)Z = 0$ and if $m \geq 4$ then $\tilde{M}$ is locally conformal flat if and only if $W(X,Y)Z = 0$. The condition of locally conformal flat holds for three dimensional Riemannian manifolds if and only if the Cotton tensor of $\tilde{M}$, which is given by

$$C(X,Y) = (\nabla_X S)Y - (\nabla_Y S)X - \frac{1}{2(m-2)}[(\nabla_X \rho)Y - (\nabla X \rho)Y],$$

vanishes identically.

The Weyl curvature tensor of real hypersurfaces M in $M_n(c)$ satisfies the relation

$$
\begin{aligned}
W(X,Y)Z &= R(X,Y)Z + \frac{1}{2n-3}[g(SX,Z)Y - g(SY,Z)X + g(X,Z)SY - g(Y,Z)SX] \\
&\quad - \frac{\rho}{2(n-1)(2n-3)}[g(X,Z)Y - g(Y,Z)X],
\end{aligned}
$$

for all X, Y, Z tangent to M, where R is the Riemannian curvature tensor, S is the Ricci tensor, ρ is the scalar curvature of M and g is the induced Riemannian metric on M. In [7] the non-existence of real hypersurfaces in $M_n(c)$ with harmonic Weyl curvature tensor, i.e., $\delta W = 0$ with δ denoting the codifferential of the exterior differential d is proved. Moreover, in [8] the classification of real hypersurfaces in $\mathbb{C}P^n$ with ξ-parallel Weyl curvature tensor, i.e., $\nabla_\xi W = 0$ is provided. Finally, in [9] real hypersurfaces in $\mathbb{C}H^n$, $n \geq 3$ satisfying the previous geometric condition are classified.

In 1959 Tachibana defined *-Ricci tensor S^* on almost Hermitian manifold. In [10] Hamada gave the definition of *-Ricci tensor S^* on real hypersurfaces in $M_n(c)$ in the following way

$$g(S^*X, Y) = \frac{1}{2} trace(Z \to R(X, \varphi Y)\varphi Z),$$

for all X, Y tangent to M and $trace$ is the sum of elements of the main diagonal of the matrix, which corresponds to the above endomorphism. He also presented *- *Einstein*, i.e., $g(S^*X, Y) = \lambda g(X, Y)$, where λ is a constant multiple of $g(X, Y)$ and provided classification of *-Einstein hypersurfaces. Ivey and Ryan in [11] extended the Hamada's work and studied the equivalence of *- Einstein condition with other geometric conditions such as the pseudo-Einstein and the pseudo-Ryan condition.

Motivated by the revious results and work we define *-Weyl curvature tensor of real hypersurfaces in the following way

$$
\begin{aligned}
W^*(X, Y)Z &= R(X, Y)Z + \frac{1}{2n-3}[g(S^*X, Z)Y - g(S^*Y, Z)X + g(X, Z)S^*Y - g(Y, Z)S^*X] \\
&\quad - \frac{\rho^*}{2(n-1)(2n-3)}[g(X, Z)Y - g(Y, Z)X],
\end{aligned}
\tag{2}
$$

for all X, Y, Z tangent to M and S^* is the *-Ricci tensor and ρ^* is the *-scalar curvature corresponding to S^* of M.

First it is examined if there are real hypersurfaces of dimension equal to or greater than three with vanishing *-Weyl curvature tensor. The following Theorem is proved

Theorem 1. *Let M be a Hopf hypersurface in $M_n(c)$, $n \geq 2$, with vanishing *-Weyl curvature tensor. Then M is an open subset of a real hypersurface of type (A) or of a Hopf hypersurface with $A\xi = 0$.*

Next it is examined if there are three-dimensional real hypersurface in $M_2(c)$ with vanishing *-Weyl curvature tensor and the following Theorem is obtained

Theorem 2. *Every real hypersurface M in $M_2(c)$ with vanishing *-Weyl curvature tensor is a Hopf hypersurface. Furthermore, M is an open subset of a real hypersurface of type (A) or of a Hopf hypersurface with $A\xi = 0$.*

The paper has the following outline: In Section 2 relations and Theorems concerning real hypersurfaces in non-flat complex space forms are provided. In Section 3 Theorems 1 and 2 are proved. Section 4 concerns discussion on the new tensor and ideas of further research and Section 5 includes the conclusions of the paper.

2. Preliminaries

The manifolds, vector fields, etc., are considered of class C^∞. We consider M to be a connected real hypersurface without boundary in $M_n(c)$ equipped with a Kahler structure (J, G) and $\overline{\nabla}$ is the

Levi-Civita connection of $M_n(c)$ and N a locally unit normal vector field on M. Then the shape operator A of M with respect to N is given by

$$\overline{\nabla}_X N = -AX.$$

and the Levi-Civita connection ∇ of the induced metric g on M satisfies

$$\overline{\nabla}_X Y = \nabla_X Y + g(AX, Y)N.$$

As mentioned in the Introduction, on M an almost contact metric structure (ϕ, ξ, η, g) is defined and the following relations are satisfied (see [12])

$$\phi^2 X = -X + \eta(X)\xi, \quad \eta(\xi) = 1, \quad g(\phi X, \phi Y) = g(X, Y) - \eta(X)\eta(Y) \tag{3}$$

for all tangent vectors X, Y to M. Relation (3) implies

$$\phi\xi = 0, \quad \eta(X) = g(X, \xi).$$

Due to the fact that the complex structure J is parallel, i.e., $\overline{\nabla}J = 0$ we have

$$(\nabla_X \phi)Y = \eta(Y)AX - g(AX, Y)\xi \quad \text{and} \quad \nabla_X \xi = \phi AX \tag{4}$$

for all X, Y tangent to M. Moreover, the ambient space is of holomorphic sectional curvature c and this results in the Gauss and Codazzi equations becoming respectively

$$R(X,Y)Z = \frac{c}{4}[g(Y,Z)X - g(X,Z)Y + g(\phi Y, Z)\phi X - g(\phi X, Z)\phi Y$$
$$-2g(\phi X, Y)\phi Z] + g(AY, Z)AX - g(AX, Z)AY, \tag{5}$$

and

$$(\nabla_X A)Y - (\nabla_Y A)X = \frac{c}{4}[\eta(X)\phi Y - \eta(Y)\phi X - 2g(\phi X, Y)\xi], \tag{6}$$

for all tangent vectors X, Y, Z to M, where R is the Riemannian curvature tensor of M.

Let P be a point of M, then the tangent space $T_P M$ is decomposed into

$$T_P M = span\{\xi\} \oplus \mathbb{D},$$

where $\mathbb{D} = \ker \eta = \{X \in T_P M : \eta(X) = 0\}$ and is called *(maximal) holomorphic distribution* (if $n \geq 3$).

The following Theorem concerns the shape operator of M and is proved by Maeda [13] in the case of $\mathbb{C}P^n, n \geq 2$, and by Ki and Suh [14] in the case of $\mathbb{C}H^n, n \geq 2$ (also Corollary 2.3 in [15]).

Theorem 3. *Let M be a Hopf hypersurface in $M_n(c)$, $n \geq 2$. Then*

(i) α *is constant.*

(ii)   *If W is a vector field which belongs to $\mathbb{D}$ such that $AW = \lambda W$, then*

$$(\lambda - \frac{\alpha}{2})A(\phi W) = (\frac{\lambda\alpha}{2} + \frac{c}{4})(\phi W). \tag{7}$$

(iii)   *If the vector field W satisfies $AW = \lambda W$ and $A(\phi W) = \nu(\phi W)$ then*

$$\lambda\nu = \frac{\alpha}{2}(\lambda + \nu) + \frac{c}{4}. \tag{8}$$

We consider M a three dimensional real hypersurface in $M_2(c)$ and P a point of M such that in the neighborhood of P relation $A\xi \neq \alpha\xi$ holds. Let U be a unit vector lying in the $span\{A\xi, \xi\}$ satisfying relation $g(U, \xi) = 0$. Then, we can consider the standard non-Hopf local orthonormal frame $\{U, \phi U, \xi\}$ in the neighborhood of P (see [16] p. 445). Therefore, the shape operator A is given by

$$A\xi = \alpha\xi + \beta U, \quad AU = \gamma U + \delta(\phi U) + \beta\xi \quad \text{and} \quad A(\phi U) = \delta U + \mu(\phi U). \tag{9}$$

The following Lemma holds for three dimensional non-Hopf real hypersurfaces in $M_2(c)$

Lemma 1. *Let M be a non-Hopf real hypersurface in $M_2(c)$. The following relations hold on M*

$$\nabla_U\xi = -\delta U + \gamma(\phi U), \qquad \nabla_{\phi U}\xi = -\mu U + \delta(\phi U), \qquad \nabla_\xi\xi = \beta(\phi U),$$
$$\nabla_U U = \kappa_1(\phi U) + \delta\xi, \qquad \nabla_{\phi U} U = \kappa_2(\phi U) + \mu\xi, \qquad \nabla_\xi U = \kappa_3(\phi U),$$
$$\nabla_U(\phi U) = -\kappa_1 U - \gamma\xi, \quad \nabla_{\phi U}(\phi U) = -\kappa_2 U - \delta\xi, \quad \nabla_\xi(\phi U) = -\kappa_3 U - \beta\xi,$$

where $\alpha, \beta, \gamma, \delta, \mu, \kappa_1, \kappa_2, \kappa_3$ are smooth functions on M and $\beta \neq 0$.

Lemma 1 is proved in page 92 [17].

The Codazzi Equation (6) for $X \in \{U, \phi U\}$ and $Y = \xi$ owing to Lemma 1 results in the following relations

$$\xi\delta = \alpha\gamma + \beta\kappa_1 + \delta^2 + \mu\kappa_3 + \frac{c}{4} - \gamma\mu - \gamma\kappa_3 - \beta^2 \tag{10}$$

$$(\phi U)\alpha = \alpha\beta + \beta\kappa_3 - 3\beta\mu \tag{11}$$

$$(\phi U)\beta = \alpha\gamma + \beta\kappa_1 + 2\delta^2 + \frac{c}{2} - 2\gamma\mu + \alpha\mu \tag{12}$$

and for $X = U$ and $Y = \phi U$

$$U\delta - (\phi U)\gamma = \mu\kappa_1 - \kappa_1\gamma - \beta\gamma - 2\delta\kappa_2 - 2\beta\mu. \tag{13}$$

In the case of three dimensional Hopf hypersurfaces we consider a point P of M and we define in the neighborhood of P a local orthonormal frame as follows: since M is a Hopf hypersurface the shape operator A restricted to the holomorphic distribution $\mathbb{D}$ has distinct eigenvalues. Thus, we choose a vector W as one of the eigenvectors fields. Moreover, due to the fact that M is three dimensional, the shape operator satisfies the following relations:

$$A\xi = \alpha\xi, \quad AW = \lambda W \quad \text{and} \quad A(\phi W) = \nu(\phi W), \tag{14}$$

and Threom 3 holds.

Finally, the following Theorem concerns the classification of real hypersurfaces in $M_n(c)$, $n \geq 2$, whose shape operator A satisfies a commuting condition. It is proved by Okumura in the case of $\mathbb{C}P^n$ (see [18]) and by Montiel and Romero in the case of $\mathbb{C}H^n$ (see [19]).

Theorem 4. *Let M be a real hypersurface of $M_n(c)$, $n \geq 2$. Then $A\phi = \phi A$, if and only if M is an open subset of a homogeneous real hypersurface of type (A).*

We mention that type (A_2) hypersurfaces do not occur in the case of three dimensional real hypersurface in $M_2(c)$.

3. Proof of Theorems 1 and 2

The *-Ricci tensor of a real hypersurface M in a non-flat complex space form is given by

$$S^*X = -[\frac{cn}{2}\phi^2 X + \phi A(\phi(AX))], \tag{15}$$

for all X tangent to M.

Let M be a Hopf hypersurface in $M_n(c)$, $n \geq 2$, with vanishing *-Weyl curvature tensor, i.e.,

$$W^*(X,Y)Z = 0. \tag{16}$$

Since M is a Hopf hypersurface ξ is an eigenvector of the shape operator relation (1) holds and relation (15) for $X = \xi$ yields $S^*\xi = 0$. Next, we consider W a unit vector field which belongs to the (maximal) holomorphic distribution such that relation $AW = \lambda W$ holds at some point $P \in M$ and relation (7) is satisfied. We have two cases:

Case I: $\alpha^2 + c \neq 0$.

In this case $\lambda \neq \frac{\alpha}{2}$ so relation (7) implies $A\phi W = \nu\phi W$ and relation (8) holds.

Relation (16) for $Z = \xi$ taking into account (2) implies

$$R(X,Y)\xi + \frac{1}{2n-3}[g(S^*X,\xi)Y - g(S^*Y,\xi)X + \eta(X)S^*Y - \eta(Y)S^*X]$$
$$- \frac{\rho^*}{2(n-1)(2n-3)}[\eta(X)Y - \eta(Y)X] = 0, \tag{17}$$

for all X, Y tangent to M.

The inner product of relation (17) for $X = W$ and $Y = \xi$ with W because of (3), (5), (15), $S^*\xi = 0$, $AW = \lambda W$ and $A(\phi W) = \nu(\phi W)$ yields

$$(\frac{c}{4} + \alpha\lambda) - \frac{1}{2n-3}(\frac{cn}{2} + \lambda\nu) + \frac{\rho^*}{2(n-1)(2n-3)} = 0. \tag{18}$$

Furthermore, the inner product of relation (17) for $X = \phi W$ and $Y = \xi$ with ϕW due to (3), (5) and (15), $S^*\xi = 0$, $AW = \lambda W$ and $A(\phi W) = \nu(\phi W)$ implies

$$(\frac{c}{4} + \alpha\nu) - \frac{1}{2n-3}(\frac{cn}{2} + \lambda\nu) + \frac{\rho^*}{2(n-1)(2n-3)} = 0. \tag{19}$$

Combination of relations (18) and (19) results in

$$\alpha(\lambda - \nu) = 0.$$

So, either $\alpha = 0$ and M is an open subset of a Hopf hypersurface with $A\xi = 0$ or $\lambda = \nu$ which implies that $A\phi = \phi A$ and because of Theorem 4 M is an open subset of a real hypersurface of type (A).

Case II: $\alpha^2 + c = 0$.

This case occurs only when the ambient space is the complex hyperbolic space $\mathbb{C}H^n$. Thus, $\alpha^2 - 4 = 0$ and this results in $\alpha = 2$. We consider W a unit vector field, which belongs to the (maximal) holomorphi distribution such that relation $AW = \lambda W$ holds at some point $P \in M$. Therefore, relation (7) due to $\alpha = 2$ and $c = -4$ implies

$$(\lambda - 1)A(\phi W) = (\lambda - 1)(\phi W).$$

First we suppose that $\lambda \neq 1$. Then the above relation implies $A(\phi W) = \phi W$. So, the inner product of relation (17) for $X = W$ and $Y = \xi$ with W because of (3), (5) and (15) for $X = \xi$ which implies $S^*\xi = 0$, $AW = \lambda W$ and $A(\phi W) = \phi W$ results in

$$(2\lambda - 1) - \frac{1}{2n-3}(\lambda - 2n) + \frac{\rho^*}{2(n-1)(2n-3)} = 0. \qquad (20)$$

Moreover, the inner product of relation (17) for $X = \phi W$ and $Y = \xi$ with ϕW due to (3), (5), (15), $S^*\xi = 0$, $AW = \lambda W$ and $A(\phi W) = \phi W$ implies

$$1 - \frac{1}{2n-3}(\lambda - 2n) + \frac{\rho^*}{2(n-1)(2n-3)} = 0. \qquad (21)$$

Combination of relations (20) and (21) yields $\lambda = 1$, which is a contradiction.

Therefore, we have $\lambda = 1$ for any vector field $W \in \mathbb{D}$ and M is an open subset of a horosphere, which is a real hypersurface of type (A) and this completes the proof of Theorem 1.

Remark 1. *Examples of Hopf hypersurfaces with $\alpha = 0$ are the following:*

- *A geodesic hypersphere of radius $r = \frac{\pi}{4}$ in $\mathbb{C}P^n$ has $\alpha = 0$.*
- *In [20,21] there are examples of Hopf hypersurfaces with $A\xi = 0$, which do not have constant principal curvatures, i.e., the eigenvalues of the shape operator corresponding to the (maximal) holomorphic distribution are not constant.*

Next we examine non-Hopf three-dimensional real hypersurfaces M in $M_2(c)$ whose *-Weyl tensor vanishes identically, i.e., relation (16) holds. We consider $\mathcal{N}$ the open subset of M such that

$$\mathcal{N} = \{P \in M : \beta \neq 0, \text{ in a neighborhood of } P\},$$

and $\{U, \phi U, \xi\}$ be the local orthonormal frame in the neighborhood of a point P defined as in Section 2. Relation (2) for $Z = \xi$ and due to $n = 2$ implies

$$R(X,Y)\xi + g(S^*X,\xi)Y - g(S^*Y,\xi)X + \eta(X)S^*Y - \eta(Y)S^*X - \frac{\rho^*}{2}[\eta(X)Y - \eta(Y)X], \qquad (22)$$

for all X, Y tangent to M. The inner product of relation (22) for $X = U$ and $Y = \xi$ with ϕU and U taking into account relations (9), (5) and (15) yields respectively

$$\alpha\delta = 0 \quad \text{and} \quad \alpha\gamma + \delta^2 + \frac{\rho^*}{2} = \frac{3c}{4} + \beta^2 + \gamma\mu. \qquad (23)$$

Moreover, the inner product of relation (22) for $X = \phi U$ and $Y = \xi$ with ϕU because of relations (9), (5) and (15) and the second of (23) results in

$$\alpha\mu = \alpha\gamma - \beta^2. \qquad (24)$$

Suppose that $\delta \neq 0$ then the first of (23) gives $\alpha = 0$. Substitution of the latter in (24) results in $\beta = 0$, which is a contradiction. Thus, relation $\delta = 0$ holds.

Relation (22) for $X = U$ and $Y = \phi U$ because of (5) implies $\mu = 0$. So, relation (24) results in $\beta^2 = \alpha\gamma$. Differentiating the latter with respect to ϕU taking into account relations (10)–(13) results in $c = 0$.

So $\mathcal{N}$ is empty and the following Proposition has been proved.

Proposition 1. *Every real hypersurface in $M_2(c)$ whose *-Weyl curvature tensor vanishes identically is a Hopf hypersurface.*

The above proposition with Theorem 1 for the case of $n = 2$ completes the proof of Theorem 2.

4. Discussion

In literature it is known that there are no Einstein real hypersurfaces in non-flat complex space forms, i.e., real hypersurfaces whose Ricci tensor satisfies relation $S = \alpha g$, where α is constant (see [15]). Therefore, new notions such as η-Einstein, i.e., the Ricci tensor satisfies relation $S = \alpha + \eta \otimes \xi$ or *-Ricci Einstein, i.e., the *-Ricci tensor satisfies $S^* = \rho^* g$, with ρ^* being constant, are introduced and the real hypersurfaces are studied with respect to the previous relations (see [10,11,15]). Thus, the next step is to introduce new tensors on real hypersurfaces in non-flat complex space forms related to the *-Ricci tensor, since there are results concerning notions and tensors related to the Ricci tensor. In this paper, we introduced the *-Weyl curvature tensor and studied real hypersurfaces in non-flat complex space forms in terms of it. Further work can be done in this direction. So, at this point some ideas for further research are mentioned:

1. it is worthwhile to study if there are non-Hopf real hypersurfaces of dimension greater than three in non-flat complex space forms with vanishing *-Weyl curvature tensor,
2. the *-Weyl curvature tensor could also be defined on real hypersurfaces in other symmetric Hermitian space forms such as the complex two-plane Grassmannians or the complex hyperbolic two-plane Grassmannians and it could be examined if there are real hypersurfaces with vanishing *-Weyl curvature tensor.

Overall, real hypersurfaces in non-flat complex space forms can be potentially applied to finding solutions of nonlinear dynamical differential equations. Ideas for research in this direction can be derived methods based on Lie algebra. For a first idea in this direction one could have a look in works (1) A Lie algebra approach to susceptible-infected-susceptible epidemics (see [22]), (2) Lie algebraic discussion for affinity based information diffusion in social networks (see [23]).

5. Conclusions

In this section we conclude the work which is presented in this paper.

*   We introduced a new type of tensor on real hypersurfaces in non-flat complex space forms by defining the *-Weyl curvature tensor on them. The new tensor is related to the *-Ricci tensor of a real hypersurface.
*   We initiated the study of real hypersurfaces in non-flat complex space forms in terms of this new tensor. The first geometric condition is that of the vanishing *-Weyl curvature tensor. The motivation for choosing this geometric condition is the existing results for Riemannian manifolds in terms of the Weyl curvature tensor. Thus, we proved two classifications Theorems. The first Theorem concerns Hopf hypersurfaces in non-flat complex space forms of dimension greater or equal to three with vanishing *-Weyl curvature tensor. The second Theorem provides a complete classification for three dimensional real hypersurfaces with vanishing *-Weyl curvature tensor.

Author Contributions: All authors contributed equally to this research and in writing the paper.

Funding: This research received no external funding.

Acknowledgments: The authors would like to thank the reviewers for their valuable comments in order to improve the paper.

Conflicts of Interest: The authors declare no conflict of interest.

References

1. Takagi, R. Real hypersurfaces in complex projective space with constant principal curvatures II. *J. Math. Soc. Jpn.* **1975**, *27*, 507–516. [CrossRef]
2. Takagi, R. Real hypersurfaces in complex projective space with constant principal curvatures. *J. Math. Soc. Jpn.* **1975**, *27*, 43–53. [CrossRef]
3. Takagi, R. On homogeneous real hypersurfaces in a complex projective space. *Osaka J. Math.* **1973**, *10*, 495–506.
4. Kimura, M. Real hypersurfaces and complex submanifolds in complex projective space. *Trans. Am. Math. Soc.* **1986**, *296*, 137–149. [CrossRef]
5. Montiel, S. Real hypersurfaces of a complex hyperbolic space. *J. Math. Soc. Jpn.* **1985**, *35*, 515–535. [CrossRef]
6. Berndt, J. Real hypersurfaces with constant principal curvatures in complex hyperbolic space. *J. Reine Angew. Math.* **1989**, *395*, 132–141. [CrossRef]
7. Ki, U.-H.; Nakagawa, H.; Suh, Y.J. Real hypersurfaces with harmonic Weyl tensor of a complex space form. *Hiroshima Math. J.* **1990**, *20*, 93–120. [CrossRef]
8. Baikousis, C.; Blair, D.E. On a type of real hypersurfaces in complex projective space. *Tsukuba J. Math.* **1996**, *20*, 505–515. [CrossRef]
9. Baikousis, C.; Suh, Y.J. Real hypersurfaces in complex hyperbolic space with parallel Weyl conformal curvature tensor. *Kyngpook Math. J.* **2000**, *40*, 173–184.
10. Hamada, T. Real hypersurfaces of complex space forms in terms of Ricci *-tensor. *Tokyo J. Math.* **2002**, *25*, 473–483. [CrossRef]
11. Ivey, T.A.; Ryan, P.J. The *-Ricci tensor for hypersurfaces in $\mathbb{C}P^n$ and $\mathbb{C}H^n$. *Tokyo J. Math.* **2011**, *34*, 445–471. [CrossRef]
12. Blair, D.E. Riemannian Geometry of Contact and Symplectic Manifolds. In *Progress in Mathematics*; Birkhauser Boston Inc.: Boston, MA, USA, 2002.
13. Maeda, Y. On real hypersurfaces of a complex projective space. *J. Math. Soc. Jpn.* **1976**, *28*, 529–540. [CrossRef]
14. Ki, U-H.; Suh, Y.J. On real hypersurfaces of a complex space form. *Math. J. Okayama Univ.* **1990**, *32*, 207–221. [CrossRef]
15. Niebergall, R.; Ryan ,P.J. Real hypersurfaces in complex space forms. *Math. Sci. Res. Inst. Publ.* **1997**, *32*, 233–305.
16. Cecil, T.E.; Ryan P.J. *Geometry of Hypersurfaces*; Springer: Berlin, Germany, 2015.
17. Panagiotidou, K.; Xenos, P.J. Real hypersurfaces in $\mathbb{C}P^2$ and $\mathbb{C}H^2$ whose structure Jacobi operator is Lie $\mathbb{D}$-parallel. *Note Mat.* **2012**, *32*, 89–99.
18. Okumura, M. On some real hypersurfaces of a complex projective space. *Trans. Am. Math. Soc.* **1975**, *212*, 355–364. [CrossRef]
19. Montiel, S.; Romero, A. On some real hypersurfaces of a complex hyperbolic space. *Geom. Dedicata* **1986**, *20*, 245–261. [CrossRef]
20. Cecil, T.E.; Ryan P.J. Focal sets and real hypersurfaces in complex projective space. *Trans. Am. Math. Soc.* **1982**, *269*, 481–499. [CrossRef]
21. Ivey, T.A.; Ryan, P.J. Hopf hypersurfaces of small Hopf principal cuurvatures in $\mathbb{C}H^2$. *Geom. Dedicata* **2009**, *141*, 147–161. [CrossRef]
22. Shang, Y. A Lie algebra approach to susceptible-infected-susceptible epidemics. *Electron. J. Differ. Equat.* **2012**, *233*, 1–7.
23. Shang, Y. Lie algebraic discussion for affinity based information diffusion in social networks. *Open Phys.* **2017**, *15*, 705–711. [CrossRef]

symmetry

MDPI

Article

Slant Curves and Contact Magnetic Curves in Sasakian Lorentzian 3-Manifolds

Ji-Eun Lee

Institute of Basic Science, Chonnam National University, Gwangju 61186, Korea; jieunlee12@naver.com

Received: 23 May 2019; Accepted: 10 June 2019; Published: 12 June 2019

Abstract: In this article, we define Lorentzian cross product in a three-dimensional almost contact Lorentzian manifold. Using a Lorentzian cross product, we prove that the ratio of κ and $\tau - 1$ is constant along a Frenet slant curve in a Sasakian Lorentzian three-manifold. Moreover, we prove that γ is a slant curve if and only if M is Sasakian for a contact magnetic curve γ in contact Lorentzian three-manifold M. As an example, we find contact magnetic curves in Lorentzian Heisenberg three-space.

Keywords: slant curves; Legendre curves; magnetic curves; Sasakian Lorentzian manifold

1. Introduction

As a generalization of Legendre curve, we defined the notion of slant curves in [1,2]. A curve in a contact three-manifold is said to be *slant* if its tangent vector field has constant angle with the Reeb vector field. For a contact Riemannian manifold, we proved that a slant curve in a Sasakian three-manifold is that its ratio of κ and $\tau - 1$ is constant. Baikoussis and Blair proved that, on a three-dimensional Sasakian manifold, the torsion of the Legendre curve is $+1$ ([3]).

A *magnetic curve* represents a trajectory of a charged particle moving on the manifold under the action of a magnetic field in [4]. A *magnetic field* on a semi-Riemannian manifold (M, g) is a closed two-form F. The *Lorentz force* of the magnetic field F is a $(1, 1)$-type tensor field Φ given by

$$g(\Phi(X), Y) = F(X, Y), \quad \forall X, Y \in \Gamma(TM). \tag{1}$$

The magnetic trajectories of F are curves γ on M that satisfy the *Lorentz equation*

$$\nabla_{\gamma'}\gamma' = \Phi(\gamma'), \tag{2}$$

where ∇ is the Levi–Civita connection of g. The Lorentz equation generalizes the equation satisfied by the geodesics of M, namely $\nabla_{\gamma'}\gamma' = 0$. Since the Lorentz force Φ is skew-symmetric, we have

$$\frac{d}{dt}g(\gamma', \gamma') = 2g(\Phi(\gamma'), \gamma') = 0,$$

that is, magnetic curve have constant speed $\mid \gamma' \mid = v_0$. When the magnetic curve $\gamma(t)$ is arc-length parameterized, it is called a *normal magnetic curve*. Cabreizo et al. studied a contact magnetic field in three-dimensional Sasakian manifold ([5]).

In this article, we define the magnetic curve γ with contact magnetic field $F_{\xi,q}$ of the length q in three-dimensional Sasakian Lorentzian manifold M^3. We call it the *contact magnetic curve* or *trajectories* of $F_{\xi,q}$.

Symmetry **2019**, *11*, 784; doi:10.3390/sym11060784   www.mdpi.com/journal/symmetry

In Section 3, we define a Lorentzian cross product in a three-dimensional almost contact Lorentzian manifold. Using the Lorentzian cross product, we prove that the ratio of κ and $\tau - 1$ is constant along a Frenet slant curve in a Sasakian Lorentzian three-manifold.

In Section 4, we prove that γ is a slant curve if and only if M is Sasakian for a contact magnetic curve γ in contact Lorentzian three-manifolds M. For example, we find contact magnetic curves in Lorentzian Heisenberg three-space.

2. Preliminaries

Contact Lorentzian Manifold

Let M be a $(2n + 1)$-dimensional differentiable manifold. M has an almost contact structure (φ, ξ, η) if it admits a tensor field φ of $(1, 1)$, a vector field ξ and a 1-form η satisfying

$$\varphi^2 = -I + \eta \otimes \xi, \ \eta(\xi) = 1. \tag{3}$$

Suppose M has an almost contact structure (φ, ξ, η). Then, $\varphi\xi = 0$ and $\eta \circ \varphi = 0$. Moreover, the endomorphism φ has rank $2n$.

If a $(2n + 1)$-dimensional smooth manifold M with almost contact structure (φ, ξ, η) admits a compatible Lorentzian metric such that

$$g(\varphi X, \varphi Y) = g(X, Y) + \eta(X)\eta(Y), \tag{4}$$

then we say M has an almost contact Lorentzian structure (η, ξ, φ, g). Setting $Y = \xi$, we have

$$\eta(X) = -g(X, \xi). \tag{5}$$

Next, if the compatible Lorentzian metric g satisfies

$$d\eta(X, Y) = g(X, \varphi Y), \tag{6}$$

then η is a contact form on M, ξ is the associated Reeb vector field, g is an associated metric and $(M, \varphi, \xi, \eta, g)$ is called a *contact Lorentzian manifold*.

For a contact Lorentzian manifold M, one may define naturally an almost complex structure J on $M \times \mathbb{R}$ by

$$J(X, f\frac{\mathrm{d}}{\mathrm{d}t}) = (\varphi X - f\xi, \eta(X)\frac{\mathrm{d}}{\mathrm{d}t}),$$

where X is a vector field tangent to M, t is the coordinate of $\mathbb{R}$ and f is a function on $M \times \mathbb{R}$. When the almost complex structure J is integrable, the contact Lorentzian manifold M is said to be *normal* or *Sasakian*. A contact Lorentzian manifold M is normal if and only if M satisfies

$$[\varphi, \varphi] + 2d\eta \otimes \xi = 0,$$

where $[\varphi, \varphi]$ is the Nijenhuis torsion of φ.

Proposition 1 ([6,7]). *An almost contact Lorentzian manifold* $(M^{2n+1}, \eta, \xi, \varphi, g)$ *is Sasakian if and only if*

$$(\nabla_X \varphi)Y = g(X, Y)\xi + \eta(Y)X. \tag{7}$$

Using the similar arguments and computations in [8], we obtain

Proposition 2 ([6,7]). *Let* $(M^{2n+1}, \eta, \xi, \varphi, g)$ *be a contact Lorentzian manifold. Then,*

$$\nabla_X \xi = \varphi X - \varphi h X. \tag{8}$$

If ξ is a killing vector field with respect to the Lorentzian metric g. Then, we have

$$\nabla_X \xi = \varphi X. \tag{9}$$

3. Slant Curves in Contact Lorentzian Three-Manifolds

Let $\gamma : I \to M^3$ be a unit speed curve in Lorentzian three-manifolds M^3 such that γ' satisfies $g(\gamma', \gamma') = \varepsilon_1 = \pm 1$. The constant ε_1 is called the *causal character* of γ. A unit speed curve γ is said to be a spacelike or timelike if its causal character is 1 or -1, respectively.

A unit speed curve γ is said to be a *Frenet curve* if $g(\gamma'', \gamma'') \neq 0$. A Frenet curve γ admits an orthonormal frame field $\{E_1 = \dot{\gamma}, E_2, E_3\}$ along γ. The constants ε_2 and ε_3 are defined by

$$g(E_i, E_i) = \varepsilon_i, \quad i = 2, 3$$

and called *second causal character* and *third causal character* of γ, respectively. Thus, $\varepsilon_1 \varepsilon_2 = -\varepsilon_3$ is satisfied. Then, the *Frenet–Serret* equations are the following ([9,10]):

$$\begin{cases} \nabla_{\dot{\gamma}} E_1 = & \varepsilon_2 \kappa E_2, \\ \nabla_{\dot{\gamma}} E_2 = -\varepsilon_1 \kappa E_1 & - \varepsilon_3 \tau E_3, \\ \nabla_{\dot{\gamma}} E_3 = & \varepsilon_2 \tau E_2, \end{cases} \tag{10}$$

where $\kappa = |\nabla_{\dot{\gamma}} \dot{\gamma}|$ is the *geodesic curvature* of γ and τ its *geodesic torsion*. The vector fields E_1, E_2 and E_3 are called tangent vector field, principal normal vector field, and binormal vector field of γ, respectively.

A Frenet curve γ is a *geodesic* if and only if $\kappa = 0$. A Frenet curve γ with constant geodesic curvature and zero geodesic torsion is called a *pseudo-circle*. A *pseudo-helix* is a Frenet curve γ whose geodesic curvature and torsion are constant.

3.1. Lorentzian Cross Product

C. Camci ([11]) defined a cross product in three-dimensional almost contact Riemannian manifolds $(\tilde{M}, \eta, \xi, \varphi, \tilde{g})$ as following:

$$X \wedge Y = -\tilde{g}(X, \varphi Y) \xi - \eta(Y) \varphi X + \eta(X) \varphi Y. \tag{11}$$

If we define the cross product $\wedge$ as Equation (11) in three-dimensional almost contact Lorentzian manifold $(M, \eta, \xi, \varphi, g)$, then

$$g(X \wedge Y, X) = 2\eta(X) g(X, \varphi Y) \neq 0.$$

In fact, we see already the cross product for a Lorentzian three-manifold as following:

Proposition 3. *Let* $\{E_1, E_2, E_3\}$ *be an orthonomal frame field in a Lorentzian three-manifold. Then,*

$$E_1 \wedge_L E_2 = \varepsilon_3 E_3, \quad E_2 \wedge_L E_3 = \varepsilon_1 E_1, \quad E_3 \wedge_L E_1 = \varepsilon_2 E_2. \tag{12}$$

Now, in three-dimensional almost contact Lorentzian manifold M^3, we define Lorentzian cross product as the following:

Symmetry **2019**, *11*, 784

Definition 1. *Let* $(M^3, \varphi, \xi, \eta, g)$ *be a three-dimensional almost contact Lorentzian manifold. We define a Lorentzian cross product* $\wedge_L$ *by*

$$X \wedge_L Y = g(X, \varphi Y)\xi - \eta(Y)\varphi X + \eta(X)\varphi Y, \tag{13}$$

where $X, Y \in TM$.

The Lorentzian cross product $\wedge_L$ has the following properties:

Proposition 4. *Let* $(M^3, \varphi, \xi, \eta, g)$ *be a three-dimensional almost contact Lorentzian manifold. Then, for all* $X, Y, Z \in TM$ *the Lorentzian cross product has the following properties:*

(1) *The Lorentzian cross product is bilinear and anti-symmetric.*
(2) $X \wedge_L Y$ *is perpendicular both of X and Y.*
(3) $X \wedge_L \varphi Y = -g(X, Y)\xi - \eta(X)Y.$
(4) $\varphi X = \xi \wedge_L X.$
(5) *Define a mixed product by* $\det(X, Y, Z) = g(X \wedge_L Y, Z)$ *Then,*

$$\det(X, Y, Z) = -g(X, \varphi Y)\eta(Z) - g(Y, \varphi Z)\eta(X) - g(Z, \varphi X)\eta(Y)$$

and $\det(X, Y, Z) = \det(Y, Z, X) = \det(Z, X, Y).$
(6) $g(X, \varphi Y)Z + g(Y, \varphi Z)X + g(Z, \varphi X)Y = -(X, Y, Z)\xi.$

Proof. (We can prove by a similar way as in [11])
(1) and (2) are trivial.
(3) using Equations (3), (5) and (13),

$$\begin{aligned} X \wedge_L \varphi Y &= g(X, -Y + \eta(Y)\xi)\xi + \eta(X)(-Y + \eta(Y)\xi) \\ &= -g(X, Y)\xi - \eta(X)Y. \end{aligned}$$

(4) by Equation (13),

$$\xi \wedge_L X = g(\xi, \varphi X)\xi - \eta(X)\varphi\xi + \eta(\xi)\varphi X = \varphi X.$$

(5) from Equations (5) and (13),

$$\begin{aligned} g(X \wedge_L Y, Z) &= g(g(X, \varphi Y)\xi - \eta(Y)\varphi X + \eta(X)\varphi Y, Z) \\ &= -g(X, \varphi Y)\eta(Z) - g(Y, \varphi Z)\eta(X) - g(Z, \varphi X)\eta(Y). \end{aligned}$$

(6) is easily obtained by (5). □

From Equations (7) and (9), we have:

Proposition 5. *Let* $(M^3, \varphi, \xi, \eta, g)$ *be a three-dimensional Sasakian Lorentzian manifold. Then, we have*

$$\nabla_Z(X \wedge_L Y) = (\nabla_Z X) \wedge_L Y + X \wedge_L (\nabla_Z Y), \tag{14}$$

for all $X, Y, Z \in TM$.

Proof. From Equation (13), we get

$$
\begin{aligned}
\nabla_Z(X \wedge_L Y) &= \nabla_Z(-g(X, \varphi Y)\xi + \eta(Y)\varphi X - \eta(X)\varphi Y) \\
&= g(\nabla_Z X, \varphi Y)\xi + g(X, (\nabla_Z \varphi)Y)\xi + g(X, \varphi \nabla_Z Y)\xi + g(X, \varphi Y)\nabla_Z \xi \\
&\quad -\eta(\nabla_Z Y)\varphi X + g(Y, \nabla_Z \xi)\varphi X + \eta(Y)(\nabla_Z \varphi)X + \eta(Y)\varphi \nabla_Z X \\
&\quad +\eta(\nabla_Z X)\varphi Y - g(X, \nabla_Z \xi)\varphi Y - \eta(X)(\nabla_Z \varphi)Y - \eta(X)\varphi \nabla_Z Y \\
&= (\nabla_Z X) \wedge_L Y + X \wedge_L (\nabla_Z Y) + P(X, Y, Z),
\end{aligned}
$$

where

$$
\begin{aligned}
P(X, Y, Z) &= g(X, (\nabla_Z \varphi)Y)\xi + g(X, \varphi Y)\nabla_Z \xi + g(Y, \nabla_Z \xi)\varphi X - \eta(Y)(\nabla_Z \varphi)X \\
&\quad -g(X, \nabla_Z \xi)\varphi Y + \eta(X)(\nabla_Z \varphi)Y.
\end{aligned}
$$

Since M is a three-dimensional Sasakian Lorentzian manifold, it satisfies Equations (7) and (9). Hence, we have

$$
P(X, Y, Z) = g(X, \varphi Y)\varphi Z + g(Y, \varphi Z)\varphi X + g(Z, \varphi X)\varphi Y.
$$

Using Equation (6) of Proposition 4, we obtain $P(X, Y, Z) = 0$ and Equation (14). □

3.2. Frenet Slant Curves

In this subsection, we study a Frenet slant curve in contact Lorentzian three-manifolds.

A curve in a contact Lorentzian three-manifold is said to be *slant* if its tangent vector field has constant angle with the Reeb vector field (i.e., $\eta(\gamma') = -g(\gamma', \xi)$ is a constant).

Since the Reeb vector field ξ is denoted by

$$
\xi = \sum_{i=1}^{3} \varepsilon_i g(\xi, E_i)E_i = -\sum_{i=1}^{3} \varepsilon_i \eta(E_i)E_i,
$$

using Equation (4) of Proposition 4 and Proposition 3, we have:

Proposition 6. *Let* $(M^3, \varphi, \xi, \eta, g)$ *be a three-dimensional almost contact Lorentzian manifold. Then, for a Frenet curve* γ *in* M^3, *we have*

$$
\begin{aligned}
\varphi E_1 &= \varepsilon_2 \varepsilon_3 (\eta(E_2)E_3 - \eta(E_3)E_2), \\
\varphi E_2 &= \varepsilon_3 \varepsilon_1 (\eta(E_3)E_1 - \eta(E_1)E_3), \\
\varphi E_3 &= \varepsilon_1 \varepsilon_2 (\eta(E_1)E_2 - \eta(E_2)E_1).
\end{aligned}
$$

By using Proposition 6, we find that differentiating $\eta(E_i)$ (for $i = 1, 2, 3$) along a Frenet curve γ

$$
\begin{aligned}
\eta(E_1)' &= \varepsilon_2 \kappa \eta(E_2) + g(E_1, \varphi h E_1), \\
\eta(E_2)' &= -\varepsilon_1 \kappa \eta(E_1) - \varepsilon_3(\tau - 1)\eta(E_3) + g(E_2, \varphi h E_1), \\
\eta(E_3)' &= \varepsilon_2(\tau - 1)\eta(E_2) + g(E_3, \varphi h E_1).
\end{aligned}
$$

Now, we assume that M^3 is a Sasakian Lorentzian manifold; then,

$$\eta(E_1)' = \varepsilon_2 \kappa \eta(E_2), \tag{15}$$

$$\eta(E_2)' = -\varepsilon_1 \kappa \eta(E_1) - \varepsilon_3(\tau - 1)\eta(E_3), \tag{16}$$

$$\eta(E_3)' = \varepsilon_2(\tau - 1)\eta(E_2). \tag{17}$$

From Equation (15), if γ is a geodesic curve, that is $\kappa = 0$, in a Sasakian Lorentzian three-manifold M^3, then γ is naturally a slant curve. Now, let us consider a non-geodesic curve γ; then, we have:

Proposition 7. *A non-geodesic Frenet curve γ in a Sasakian Lorentzian three-manifold M^3 is slant curve if and only if $\eta(E_2) = 0$.*

From Equations (15) and (17) and Proposition 7, we get that $\eta(E_1)$ and $\eta(E_3)$ are constants. Hence, using Equation (16), we obtain:

Theorem 1. *The ratio of κ and $\tau - 1$ is a constant along a non-geodesic Frenet slant curve in a Sasakian Lorentzian three-manifold M^3.*

Next, let us consider a Legendre curve γ as a spacelike curve with spacelike normal vector. For a Legendre curve γ, $\eta(\gamma') = \eta(E_1) = 0$, $\eta(E_2) = 0$ and $\eta(E_3)$ is a constant. Hence, using Equation (16), we have:

Corollary 1. *Let M be a three-dimensional Sasakian Lorentzian manifold $(M^3, \eta, \xi, \varphi, g)$. Then, the torsion of a Legendre curve is 1.*

From this, we see that the ratio of κ and $\tau - 1$ is a constant along non-geodesic Frenet slant curve containing Legendre curve.

3.3. Null Slant Curves

In this section, let us consider a null curve γ that has a null tangent vector field $g(\gamma', \gamma') = 0$ and γ is not a geodesic (i.e., $g(\nabla_{\gamma'}\gamma', \nabla_{\gamma'}\gamma') \neq 0$). We take a parameterization of γ such that $g(\nabla_{\gamma'}\gamma', \nabla_{\gamma'}\gamma') = 1$. Then, Duggal, K.L. and Jin, D.H ([12]) proved that there exists only one Cartan frame $\{T, N, W\}$ and the function τ along γ whose Cartan equations are

$$\nabla_T T = N, \quad \nabla_T W = \tau N, \quad \nabla_T N = -\tau T - W,$$

where

$$T = \gamma', \quad N = \nabla_T T, \quad \tau = \frac{1}{2}g(\nabla_T N, \nabla_T N), \quad W = -\nabla_T N - \tau T. \tag{18}$$

Hence,

$$g(T, W) = g(N, N) = 1, \quad g(T, T) = g(T, N) = g(W, W) = g(W, N) = 0.$$

For a null Legendre curve γ, we easily prove that γ is geodesic. Hence, we suppose that γ is non-geodesic; then, we have:

Theorem 2. *Let γ be a non-geodesic null slant curve in a Sasakian Lorentzian three-manifold. We assume that $\kappa = 1$, then we have*

$$N = \pm \frac{1}{a}\varphi\gamma', \quad \tau = \frac{1}{2a^2} \mp 1, \quad W = \frac{1}{2a^2}\gamma' - \frac{1}{a}\xi, \tag{19}$$

where $a = \eta(\gamma')$ is non-zero constant.

Proof. Let $\varphi T = lT + mN + nW$ for some l, m, n. We find $l = g(\varphi T, T) = 0$, then $\varphi T = mN + nW$. From this, we get

$$g(\varphi T, \varphi T) = m^2 = a^2 \quad and \quad 0 = g(\varphi T, \xi) = n(a\tau + m).$$

Hence, $m = \pm a$ and $n = 0$ or $m = -a\tau$.

If $n = 0$, then $N = \frac{1}{m}\varphi T = \pm\frac{1}{a}\varphi T$. Using the Cartan equation, we find that $\tau = \frac{1}{2a^2} \mp 1$ and $W = \frac{1}{2a^2}\gamma' - \frac{1}{a}\xi$.

Next, if $n \neq 0$ and $m = -a\tau$ then since γ is a slant curve, differentiating $g(\varphi T, N) = m = \pm a$, we have $n = g(\varphi T, W) = 0$, which gives a contradiction. $\square$

From the second equation of Equation (19), we have:

Remark 1. *Let γ be a non-geodesic null slant curve in a Sasakian Lorentzian three-manifold. We assume that $\kappa = 1$ then τ is constant such that $\tau = \frac{1}{2a^2} \mp 1$.*

4. Contact Magnetic Curves

In a three-dimensional Sasakian Lorentzian manifold M^3, the Reeb vector field ξ is Killing. By Equation (6), the 2-form Φ is $d\eta$, that is $d\eta(X, Y) = g(X, \varphi Y)$, for all $X, Y \in \Gamma(TM)$.

Let $\gamma : I \to M$ be a smooth curve on a contact Lorentzian manifold $(M, \varphi, \xi, \eta, g)$. Then, we define a magnetic field on M by

$$F_{\xi,q}(X, Y) = -qd\eta(X, Y),$$

where $X, Y \in \mathbb{X}(M)$ and q is a non-zero constant. We call $F_{\xi,q}$ the *contact magnetic field* with strength q.

Using Equations (1), (4) and (6) we get $\Phi(X) = q\varphi X$. Hence, from Equation (2) the Lorentz equation is

$$\nabla_{\gamma'}\gamma' = q\varphi\gamma'. \tag{20}$$

This is the generalized equation of geodesics under arc length parameterization, that is $\nabla_{\gamma'}\gamma' = 0$. For $q = 0$, we find that the contact magnetic field vanishes identically and the magnetic curves are geodesics of M. The solutions of Equation (20) are called *contact magnetic curve* or *trajectories* of $F_{\xi,q}$.

By using Equations (8) and (20), differentiating $g(\xi, \gamma')$ along a contact magnetic curve γ in contact Lorentzian three-manifold

$$\begin{aligned}
\frac{d}{dt}g(\xi, \gamma') &= g(\nabla_{\gamma'}\xi, \gamma') + g(\xi, \nabla_{\gamma'}\gamma') \\
&= g(\varphi\gamma' - \varphi h\gamma', \gamma') + g(\xi, q\varphi\gamma') \\
&= -g(\varphi h\gamma', \gamma').
\end{aligned}$$

Hence, we have:

Theorem 3. *Let γ be a contact magnetic curve in a contact Lorentzian three-manifold M. γ is a slant curve if and only if M is Sasakian.*

Next, we find the curvature κ and torsion τ along non-geodesic Frenet contact magnetic curves γ. We suppose that $\eta(E_1) = a$, for a constant a. Then, using Equations (4), (10) and (20), we get

$$\varepsilon_2 \kappa^2 = q^2 g(\varphi \gamma', \varphi \gamma') = q^2(\varepsilon_1 + a^2).$$

Hence, we find that γ has a constant curvature

$$\kappa = \mid q \mid \sqrt{\varepsilon_2(\varepsilon_1 + a^2)}, \tag{21}$$

and, from Equations (10), (20) and (21), the binormal vector field

$$E_2 = \frac{q}{\varepsilon_2 \kappa} \varphi \gamma' = \frac{\delta \varepsilon_2}{\sqrt{\varepsilon_2(\varepsilon_1 + a^2)}} \varphi \gamma', \tag{22}$$

where $\delta = q/ \mid q \mid$.

Using Proposition 3 and Equation (22), the binormal E_3 is computed as

$$
\begin{aligned}
\varepsilon_3 E_3 &= E_1 \wedge_L E_2 \\
&= \gamma' \wedge_L \left(\frac{\delta \varepsilon_2}{\sqrt{\varepsilon_2(\varepsilon_1 + a^2)}} \varphi \gamma' \right) \\
&= -\frac{\delta \varepsilon_2}{\sqrt{\varepsilon_2(\varepsilon_1 + a^2)}} (\varepsilon_1 \xi + a \gamma').
\end{aligned}
$$

Differentiating binormal vector field E_3, we have

$$
\begin{aligned}
\nabla_{\gamma'} E_3 &= -\frac{\delta \varepsilon_2 \varepsilon_3}{\sqrt{\varepsilon_2(\varepsilon_1 + a^2)}} \nabla_{\gamma'}(\varepsilon_1 \xi + a \gamma') \\
&= -\frac{\delta \varepsilon_2 \varepsilon_3}{\sqrt{\varepsilon_2(\varepsilon_1 + a^2)}} (\varepsilon_1 + qa) \varphi \gamma'.
\end{aligned} \tag{23}
$$

On the other hand, by Equation (10), we have

$$\nabla_{\gamma'} E_3 = \varepsilon_2 \tau E_2 = \tau \frac{\delta \varphi \gamma'}{\sqrt{\varepsilon_2(\varepsilon_1 + a^2)}}. \tag{24}$$

From Equations (23) and (24), since $\varepsilon_1 \varepsilon_2 \varepsilon_3 = -1$, we obtain

$$\tau = 1 + \varepsilon_1 qa. \tag{25}$$

Moreover, if γ is a non-geodesic curve, then

$$\frac{\tau - 1}{\kappa} = \frac{\delta \varepsilon_1 a}{\sqrt{\varepsilon_2(\varepsilon_1 + a^2)}}.$$

Therefore, we obtain:

Theorem 4. *Let γ be a non-geodesic Frenet curve in a Sasakian Lorentzian three-manifold M. If γ is a contact magnetic curve, then it is slant pseudo-helix with curvature $\kappa = \mid q \mid \sqrt{\varepsilon_2(\varepsilon_1 + a^2)}$ and torsion $\tau = 1 + \varepsilon_1 qa$. Moreover, the ratio of κ and $\tau - 1$ is a constant.*

Since a Legendre curve is a spacelike curve with spacelike normal vector field and $\eta(\gamma') = a = 0$, we assume that γ is a Legendre curve and we have:

Corollary 2. *Let γ be a non-geodesic Legendre curve in a Sasakian Lorentzian three-manifold M. If γ is a contact magnetic curve, then it is Legendre pseudo-helix with curvature $\kappa = |q|$ and torsion $\tau = 1$.*

Now, from the geodesic curvature in Equation (21), if $\varepsilon_1 = 1$, then $\eta(\gamma') = a$ and $1 \leq 1 + a^2$, and we have $\varepsilon_2 = 1$. Moreover, using $\varepsilon_3 = -\varepsilon_1 \cdot \varepsilon_2$, we obtain $\varepsilon_3 = -1$. Next, if $\varepsilon_1 = -1$, then $\eta(\gamma') = a = \cosh \alpha_0$. Since γ is a geodesic for $a = \cosh \alpha_0 = 1$, we assume that γ is non-geodesic, and we get $a^2 > 1$. Hence, $-1 + a^2 > 0$ and we get $\varepsilon_2 = \varepsilon_3 = 1$. Therefore, we obtain:

Theorem 5. *Let γ be a non-geodesic Frenet curve in a Sasakian Lorentzian three-manifold M. If γ is a contact magnetic curve. then γ is one of the following:*

(i)   *a spacelike curve with spacelike normal vector field; or*
(ii) *a timelike curve.*

Moreover, we have:

Corollary 3. *Let γ be a non-geodesic Frenet curve in a Sasakian Lorentzian three-manifold M. If γ is a contact magnetic curve, then there does not exist a spacelike curve with timelike normal vector field.*

In a similar with a Frenet curve, we study null contact magnetic curves in a Sasakian Lorentzian three-manifold M. Hence, we find that there exist a null contact magnetic curve with $q = \pm a$ and same the result with Theorem 2.

Example

The Heisenberg group $\mathbb{H}_3$ is a Lie group which is diffeomorphic to $\mathbb{R}^3$ and the group operation is defined by

$$(x, y, z) * (\overline{x}, \overline{y}, \overline{z}) = (x + \overline{x}, y + \overline{y}, z + \overline{z} + \frac{x\overline{y}}{2} - \frac{\overline{x}y}{2}).$$

The mapping

$$\mathbb{H}_3 \to \left\{ \begin{pmatrix} 1 & a & b \\ 0 & 1 & c \\ 0 & 0 & 1 \end{pmatrix} \middle| a, b, c \in \mathbb{R} \right\} : (x, y, z) \mapsto \begin{pmatrix} 1 & x & z + \frac{xy}{2} \\ 0 & 1 & y \\ 0 & 0 & 1 \end{pmatrix}$$

is an isomorphism between $\mathbb{H}_3$ and a subgroup of $GL(3, \mathbb{R})$.

Now, we take the contact form

$$\eta = dz + (y dx - x dy).$$

Then, the characteristic vector field of η is $\xi = \frac{\partial}{\partial z}$.

Now, we equip the Lorentzian metric as following:

$$g = dx^2 + dy^2 - (dz + (y dx - x dy))^2.$$

We take a left-invariant Lorentzian orthonormal frame field (e_1, e_2, e_3) on $(\mathbb{H}_3, g)$:

$$e_1 = \frac{\partial}{\partial x} - y\frac{\partial}{\partial z}, \ e_2 = \frac{\partial}{\partial y} + x\frac{\partial}{\partial z}, \ e_3 = \frac{\partial}{\partial z},$$

and the commutative relations are derived as follows:

$$[e_1, e_2] = 2e_3, \ [e_2, e_3] = [e_3, e_1] = 0.$$

Then, the endomorphism field φ is defined by

$$\varphi e_1 = e_2, \ \varphi e_2 = -e_1, \ \varphi e_3 = 0.$$

The Levi–Civita connection ∇ of $(\mathbb{H}_3, g)$ is described as

$$\nabla_{e_1} e_1 = \nabla_{e_2} e_2 = \nabla_{e_3} e_3 = 0, \ \nabla_{e_1} e_2 = e_3 = -\nabla_{e_2} e_1, \tag{26}$$
$$\nabla_{e_2} e_3 = -e_1 = \nabla_{e_3} e_2, \ \nabla_{e_3} e_1 = e_2 = \nabla_{e_1} e_3.$$

The contact form η satisfies $d\eta(X, Y) = g(X, \varphi Y)$. Moreover, the structure (η, ξ, φ, g) is Sasakian. The Riemannian curvature tensor R of $(\mathbb{H}_3, g)$ is given by

$$\begin{aligned}
R(e_1, e_2)e_1 &= 3e_2, & R(e_1, e_2)e_2 &= -3e_1, \\
R(e_2, e_3)e_2 &= -e_3, & R(e_2, e_3)e_3 &= -e_2, \\
R(e_3, e_1)e_3 &= e_1, & R(e_3, e_1)e_1 &= e_3,
\end{aligned}$$

and the other components are zero.

The sectional curvature is given by [6]

$$K(\xi, e_i) = -R(\xi, e_i, \xi, e_i) = -1, \ for \ i = 1, 2,$$

and

$$K(e_1, e_2) = R(e_1, e_2, e_1, e_2) = 3.$$

Thus, we see that the Lorentzian Heisenberg space $(\mathbb{H}_3, g)$ is the Lorentzian Sasakian space forms with constant holomorphic sectional curvature $\mu = 3$.

Let γ be a Frenet slant curve in Lorentzian Heisenberg space $(\mathbb{H}_3, g)$ parameterized by arc-length. Then, the tangent vector field has the form

$$T = \gamma' = \sqrt{\varepsilon_1 + a^2} \cos\beta e_1 + \sqrt{\varepsilon_1 + a^2} \sin\beta e_2 + a e_3, \tag{27}$$

where $a = constant$, $\beta = \beta(s)$. Using Equation (26), we get

$$\nabla_{\gamma'}\gamma' = \sqrt{\varepsilon_1 + a^2}(\beta' + 2a)(-\sin\beta e_1 + \cos\beta e_2). \tag{28}$$

Since γ is a non-geodesic, we may assume that $\kappa = \sqrt{\varepsilon_1 + a^2}(\beta' + 2a) > 0$ without loss of generality. Then, the normal vector field

$$N = -\sin\beta e_1 + \cos\beta e_2.$$

The binormal vector field $\varepsilon_3 B = T \wedge_L N = -a \cos \beta e_1 - a \sin \beta e_2 - \sqrt{\varepsilon_1 + a^2} e_3$. From Theorem 5, we see that $\varepsilon_2 = 1$, thus we have $\varepsilon_3 = -\varepsilon_1$. Hence,

$$B = \varepsilon_1 (a \cos \beta e_1 + a \sin \beta e_2 + \sqrt{\varepsilon_1 + a^2} e_3).$$

Using the Frenet–Serret Equation (10), we have

Lemma 1. *Let γ be a Frenet slant curve in Lorentzian Heisenberg space $(\mathbb{H}_3, g)$ parameterized by arc-length. Then, γ admits an orthonormal frame field $\{T, N, B\}$ along γ and*

$$\kappa = \sqrt{\varepsilon_1 + a^2}(\beta' + 2a), \tag{29}$$
$$\tau = 1 + \varepsilon_1 a(\beta' + 2a).$$

Next, if γ is a null slant curve in the Lorentzian Heisenberg space $(\mathbb{H}_3, g)$, then the tangent vector field has the form

$$T = \gamma' = a \cos \beta e_1 + a \sin \beta e_2 + a e_3, \tag{30}$$

where $a = constant$, $\beta = \beta(s)$. Using Equation (26), we get

$$\nabla_{\gamma'} \gamma' = a(\beta' + 2a)(-\sin \beta e_1 + \cos \beta e_2). \tag{31}$$

Since γ is non-geodesic, using Equation (18) we have $\mid a(\beta' + 2a) \mid = 1$ and

$$N = -\sin \beta e_1 + \cos \beta e_2.$$

Differentiating N, we get

$$\nabla_{\gamma'} N = -(\beta' + a) \cos \beta e_1 - (\beta' + a) \sin \beta e_2 + a e_3.$$

From Equation (18), $\tau = \frac{1}{2} g(\nabla_{\gamma'} N, \nabla_{\gamma'} N) = \frac{1}{2}(\beta')^2 + a\beta'$. Since $W = -\nabla_{\gamma'} N - \tau T$, we have

$$W = \{-\frac{1}{2}(\beta')^2 + (\frac{1}{a} - a)\beta' + 1\}T - (\beta' + 2a)\xi = \frac{1}{2a}(\cos \beta e_1 + \sin \beta e_2 - e_3).$$

Therefore, we have

Lemma 2. *Let γ be a non-geodesic null slant curve in the Lorentzian Heisenberg space $(\mathbb{H}_3, g)$. We assume that $\kappa = \mid a(\beta' + 2a) \mid = 1$. Then, its torsion is constant such that $\tau = \frac{1}{2a^2} \mp 1$.*

Let $\gamma(s) = (x(s), y(s), z(s))$ be a curve in Lorentzian Heisenberg space $(\mathbb{H}_3, g)$. Then, the tangent vector field γ' of γ is

$$\gamma' = \left(\frac{dx}{ds}, \frac{dy}{ds}, \frac{dz}{ds} \right) = \frac{dx}{ds} \frac{\partial}{\partial x} + \frac{dy}{ds} \frac{\partial}{\partial y} + \frac{dz}{ds} \frac{\partial}{\partial z}.$$

Using the relations:

$$\frac{\partial}{\partial x} = e_1 + y e_3, \quad \frac{\partial}{\partial y} = e_2 - x e_3, \quad \frac{\partial}{\partial z} = e_3,$$

if γ is a slant curve in $(\mathbb{H}_3, g)$, then from Equation (27) the system of differential equations for γ is given by

$$\frac{dx}{ds}(s) = \sqrt{\varepsilon_1 + a^2} \cos \beta(s), \tag{32}$$

$$\frac{dy}{ds}(s) = \sqrt{\varepsilon_1 + a^2} \sin \beta(s), \tag{33}$$

$$\frac{dz}{ds}(s) = a + \sqrt{\varepsilon_1 + a^2}(x(s) \sin \beta(s) - y(s) \cos \beta(s)).$$

Now, we construct a magnetic curve γ (containing Frenet and null curve) in the Lorentzian Heisenberg space $(\mathbb{H}_3, g)$. From Equations (20) and (28), we have:

Proposition 8. *Let $\gamma : I \to (\mathbb{H}_3, g)$ be a magnetic curve parameterized by arc-length in the Lorentzian Heisenberg space $(\mathbb{H}_3, g)$. Then,*

$$\beta' = q - 2a, \quad for\ a = \eta(\gamma').$$

Namely, β' is a constant, e.g., A, hence $\beta(s) = As + b$, $b \in \mathbb{R}$. If γ is a null curve, then $q = \pm\frac{1}{a}$. Finally, from Equations (32) and (33), we have the following result:

Theorem 6. *Let $\gamma : I \to (\mathbb{H}_3, g)$ be a non-geodesic curve parameterized by arc-length s in the Lorentzian Heisenberg group $(\mathbb{H}_3, g)$. If γ is a contact magnetic curve, then the parametric equations of γ are given by*

$$\begin{cases} x(s) = \frac{1}{A}\sqrt{\varepsilon_1 + a^2} \sin(As + b) + x_0, \\ y(s) = -\frac{1}{A}\sqrt{\varepsilon_1 + a^2} \cos(As + b) + y_0, \\ z(s) = \{a + \frac{\varepsilon_1 + a^2}{A}\}s - \frac{\sqrt{\varepsilon_1 + a^2}}{A}\{x_0 \cos(As + b) + y_0 \sin(As + b)\} + z_0, \end{cases}$$

where b, x_0, y_0, z_0 are constants. If $\varepsilon_1 = 0$ then γ is a null curve.

In particular, for a Frenet Legendre curve γ, we get $\beta' = q = A$.

Funding: The author was supported by Basic Science Research Program through the National Research Foundation of Korea (NRF) funded by the Ministry of Education, Science and Technology (NRF-2019R1I1A1A01043457).

Acknowledgments: The author would like to thank the reviewers for their valuable comments on this paper to improve the quality.

Conflicts of Interest: The author declares no conflict of interest.

References

1. Cho, J.T.; Inoguchi, J.; Lee, J.-E. On slant curves in Sasakian 3-manifolds. *Bull. Aust. Math. Soc.* **2006**, *74*, 359–367. [CrossRef]
2. Inoguchi, J.; Lee, J.-E. On slant curves in normal almost contact metric 3-manifolds. *Beiträge Algebra Geom.* **2014**, *55*, 603–620. [CrossRef]
3. Baikoussis, C.; Blair, D.E. On Legendre curves in contact 3-manifolds. *Geom. Dedic.* **1994**, *49*, 135–142. [CrossRef]
4. Barros, M.; Cabrerizo, J.L.; Fernandez, M.; Romero, A. The Gauss-Landau-Hall problem on Riemannian surfaces. *J. Math. Phys.* **2005**, *46*, 1–15. [CrossRef]
5. Cabrerizo, J.L.; Fernandez, M.; Gomez, J.S. The contact Magnetic flow in 3D Sasakian manifolds. *J. Phys. A Math. Theor.* **2009**, *42*, 195201. [CrossRef]
6. Calvaruso, G. Contact Lorentzian manifolds. *Differ. Geom. Appl.* **2011**, *29*, 541–551. [CrossRef]
7. Calvaruso, G.; Perrone, D. Contact pseudo-metric manifolds. *Differ. Geom. Appl.* **2010**, *28*, 615–634. [CrossRef]

8. Blair, D.E. Riemannian Geometry of Contact and Symplectic Manifolds. In *Progress in Mathematics 203*; Birkhäuser: Chicago, IL, USA, 2002.

9. Ferrandez, A. Riemannian Versus Lorentzian submanifolds, some open problems. In Proceedings of the Workshop on Recent Topics in Differential Geometry, Santiago de Compostela, Spain, 16–19 July 1998; pp. 109–130.

10. Inoguchi, J. Biharmonic curves in Minkowki 3-space. *Int. J. Math. Math. Sci.* **2003**, *21*, 1365–1368. [CrossRef]

11. Camci, C. Extended cross product in a 3-dimensional almost contact metric manifold with applications to curve theory. *Turk. J. Math.* **2011**, *35*, 1–14.

12. Duggal, K.L.; Jin, D.H. *Null Curves and Hypersurfaces of Semi-Riemannian Manifolds*; World Scientific Publishing: Singapore, 2007.

symmetry

MDPI

Article

The Existence of Two Homogeneous Geodesics in Finsler Geometry

Zdeněk Dušek

Institute of Technology and Business in České Budějovice, Okružní 517/10,
370 01 České Budějovice, Czech Republic; zdusek@mail.vstecb.cz

Received: 31 May 2019; Accepted: 24 June 2019; Published: 1 July 2019

Abstract: The existence of a homogeneous geodesic in homogeneous Finsler manifolds was positively answered in previous papers. However, the result is not optimal. In the present paper, this result is refined and the existence of at least two homogeneous geodesics in any homogeneous Finsler manifold is proved. In a previous paper, examples of Randers metrics which admit just two homogeneous geodesics were constructed, which shows that the present result is the best possible.

Keywords: homogeneous manifold; homogeneous Finsler space; homogeneous geodesic

MSC: 53C22; 53C60; 53C30

1. Introduction

Homogeneous spaces are a natural generalization of symmetric spaces and they keep many of their nice properties. One of them is the existence of a transitive group of transformations, which are sometimes called symmetries. The importance of geodesic curves is well known in mathematics and also in physics and homogeneous geodesics are, moreover, orbits of these symmetries. In physics, they are related with relative equilibria. In Riemannian geometry, homogeneous geodesics were studied by many authors and many results were obtained, see the recent survey paper [1] by the author. In recent years, homogeneous geodesics attained interest in Finsler geometry. In the present paper, we shall focus on the existence of homogeneous geodesics in homogeneous Finsler manifolds and on an interesting phenomenon related with nonreversibility of general Finsler metrics and consequent nonreversibility of homogeneous geodesics.

The existence of at least one homogeneous geodesic in arbitrary homogeneous Riemannian manifold was proved by O. Kowalski and J. Szenthe in [2]. In the papers [3,4], it was proved that this result is optimal, namely, examples of homogeneous Riemannian metrics on solvable Lie groups were constructed which admit just one homogeneous geodesic through any point. Generalization of this existence result to pseudo-Riemannian geometry was proved by the author using a different approach in the broader context of homogeneous affine manifolds in [5]. This affine approach was used by the author also in [6] to prove that an even-dimensional Lorentzian manifold admits a light-like homogeneous geodesic.

Generalization of this existence result to Finsler geometry was proved in the series of papers [7] by Z. Yan and S. Deng for Randers metrics, [8] by the author for odd-dimensional Finsler metrics, [9] by the author for Berwald or reversible Finsler metrics, [10] by Z. Yan and L. Huang in general. In this last paper, an original approach by O. Kowalski and J. Szenthe is modified and a purely Finslerian construction is used. However, due to the nonreversibility of general Finsler metrics, it was conjectured by the author in [11] that the result and its proofs in the nonreversible situation are not optimal. In comparison with Riemannian geometry, the situation is rather delicate. In the context of Finsler geometry, the trajectory of the unique homogeneous geodesic in a Riemannian manifold should be regarded as two geodesics—they have the same trajectory, their initial vectors are X and $-X$ and they have opposite parametrizations. For a general homogeneous Finsler manifold, the initial vectors of the two homogeneous geodesics may be non-opposite. In the paper [11], examples of invariant Randers metrics which admit just two homogeneous geodesics were constructed. The initial vectors of these geodesics are $X + Y$ and $-X + Y$, for certain vectors $X, Y \in T_p M$.

In the present paper, the mentioned proofs are revised and refined. The complete and selfcontained proof of the existence of two homogeneous geodesics through an arbitrary point in arbitrary homogeneous Finsler manifold is given. Some constructions from [2,10,12] are used.

2. Basic Settings

A Minkowski norm on the vector space $\mathbb{V}$ is a nonnegative function $F : \mathbb{V} \to \mathbb{R}$ which is smooth on $\mathbb{V} \setminus \{0\}$, positively homogeneous ($F(\lambda y) = \lambda F(y)$ for any $\lambda > 0$) and whose Hessian $g_{ij} = (\frac{1}{2} F^2)_{y^i y^j}$ is positively definite on $\mathbb{V} \setminus \{0\}$. Variables (y^i) are the components of a vector $y \in \mathbb{V}$ with respect to a basis B of $\mathbb{V}$ and putting y^i to a subscript refers to the partial derivative. The pair $(\mathbb{V}, F)$ is called a Minkowski space. The tensor g_y whose components are $g_{ij}(y)$ is the fundamental tensor. We recall the well known formulas

$$
\begin{aligned}
g_y(y, u) &= \frac{1}{2} \frac{dF^2(y + su)}{ds}\Big|_{s=0}, \qquad \forall y, u \in \mathbb{V}, \\
g_y(y, y) &= F^2(y), \qquad \forall y \in \mathbb{V}.
\end{aligned}
\tag{1}
$$

A Finsler metric on a differentiable manifold M is a function F on TM which is differentiable on $TM \setminus \{0\}$ and such that its restriction to any tangent space $T_x M$ is a Minkowski norm. The pair (M, F) is called a Finsler manifold. On a Finsler manifold, functions g_{ij} depend differentiably on $x \in M$ and on $o \neq y \in T_x M$.

Let M be a Finsler manifold (M, F). If some connected Lie group G acts transitively on M by isometries, then M is called a homogeneous manifold. We remark that a homogeneous manifold (M, F) may admit more presentations as a homogeneous space in the form G/H, corresponding to various transitive isometry groups.

Homogeneous manifold M can be identified with the homogeneous space G/H. Here H is the isotropy group of the origin $p \in M$. A homogeneous Finsler space $(G/H, F)$ is a reductive homogeneous space in the following sense: Denote by $\mathfrak{g}$ and $\mathfrak{h}$ the Lie algebras of the groups G and H, respectively, and consider the representation $\mathrm{Ad} \colon H \times \mathfrak{g} \to \mathfrak{g}$ of H on $\mathfrak{g}$. There exists a reductive decomposition $\mathfrak{g} = \mathfrak{m} + \mathfrak{h}$ where $\mathfrak{m} \subset \mathfrak{g}$ is a vector subspace with the property $\mathrm{Ad}(H)(\mathfrak{m}) \subset \mathfrak{m}$. For a fixed reductive decomposition $\mathfrak{g} = \mathfrak{m} + \mathfrak{h}$ it is natural to identify $\mathfrak{m} \subset \mathfrak{g} = T_e G$ with the tangent space $T_p M$ via the projection $\pi \colon G \to G/H = M$. Using this identification, from the Minkovski norm and its fundamental tensor on $T_p M$, we obtain the $\mathrm{Ad}(H)$-invariant Minkowski norm and the $\mathrm{Ad}(H)$-invariant fundamental tensor on $\mathfrak{m}$.

We further recall the slit tangent bundle TM_0, which is defined as $TM_0 = TM \setminus \{0\}$. Using the restriction of the projection $\pi \colon TM \to M$ to TM_0, we construct the pullback vector bundle $\pi^* TM$ over TM_0. The Chern connection is the unique linear connection on $\pi^* TM$ which is torsion free and almost g-compatible. See some monograph, for example [13] by D. Bao, S.-S. Chern and Z. Shen or [14] by S. Deng for details. Using the Chern connection, the derivative along a curve $\gamma(t)$ can be defined. A regular differentiable curve γ with tangent vector field T is a *geodesic* if it holds $D_T(\frac{T}{F(T)}) = 0$. In particular, for a geodesic of constant speed it holds $D_T T = 0$.

A geodesic $\gamma(s)$ through the point p is homogeneous if it is an orbit of a one-parameter group of isometries. Explicitly, if there exists a nonzero vector $X \in \mathfrak{g}$ such that $\gamma(t) = \exp(tX)(p)$ for all $t \in \mathbb{R}$. Such a vector X is called a geodesic vector. Geodesic vectors are characterized by the geodesic lemma, proved in Riemannian geometry by O. Kowalski and L. Vanhecke in [15] and generalized to Finsler geometry by D. Latifi in [16].

Lemma 1 ([16]). *Let $(G/H, F)$ be a homogeneous Finsler space with a reductive decomposition $\mathfrak{g} = \mathfrak{m} + \mathfrak{h}$. A nonzero vector $y \in \mathfrak{g}$ is geodesic if and only if it holds*

$$g_{y_{\mathfrak{m}}}(y_{\mathfrak{m}}, [y, u]_{\mathfrak{m}}) = 0 \qquad \forall u \in \mathfrak{m},$$

where the subscript $\mathfrak{m}$ indicates the projection of a vector from $\mathfrak{g}$ to $\mathfrak{m}$.

3. The Main Result

Theorem 1. *Let (M, F) be a homogeneous Finsler manifold. There exist at least two homogeneous geodesics through arbitrary point $p \in M$.*

Proof. Let G be a transitive isometry group of M and let H be the isotropy group of a fixed point $p \in M$. We express M as the homogeneous space $M = G/H$. Let K be the Killing form on G and let $\mathrm{Rad}(K)$ be the null space of K. We choose $\mathfrak{m} = \mathfrak{h}^{\perp}$ with respect to K. The decomposition in $\mathrm{Ad}(H)$-invariant and the Finsler metric induces the invariant Minkowski norm and its fundamental tensor on $\mathfrak{m}$. We shall denote these again by F and g. The Killing form K is negatively semidefinite on $\mathfrak{g}$ and negatively definite on $\mathfrak{h}$, because H is compact. Hence, $\mathrm{Rad}(K) \subseteq \mathfrak{m}$. We shall distinguish the two cases:

(Case 1) $\mathrm{Rad}(K) = \mathfrak{m}$: we chose a hyperplane $W \subset \mathfrak{m}$ such that $[\mathfrak{m}, \mathfrak{m}] \subset W$. We used the construction and notation from [12] to show that there exist two vectors $n_1, n_2 \in \mathfrak{m}$ such that

$$g_{n_i}(n_i, w) = 0 \qquad \forall w \in W, \qquad i = 1, 2.$$

Consider an arbitrary fixed vector $v \notin W$. The function $\phi(w) := F(v - w)$ defined on W attains its minimum m at a unique point $w_0 \in W$. We put

$$n_1 = \frac{v - w_0}{m}.$$

It can be proved that the definition of the vector n_1 does not depend on the choice of the vector v on the same side of the hyperplane W. If we start with a vector v on the other side of the hyperplane W, the same construction leads to the vector n_2 on the other side of the hyperplane W and it is in general not opposite to n_1, unless F is reversible. We shall now write n for any of the two vectors n_1, n_2. For an arbitrary fixed vector $w \in W$, the equality

$$F^2(n + tw) = \frac{1}{m^2} F^2(v - w_0 + tmw) = \frac{1}{m^2} \phi^2(w_0 - tmw),$$

shows that the function $F^2(n + tw)$ attains its minimum at $t = 0$ and hence, using Formula (1), it holds

$$0 = \frac{1}{2} \frac{d}{dt} F^2(n + tw)\big|_{t=0} = g_n(n, w), \qquad \forall w \in W,$$

which is the desired property. In particular, it is satisfied for any $w \in [\mathfrak{m}, \mathfrak{m}] \subset W$. We obtain immediately, using Lemma 1, that n_1 and n_2 are geodesic vectors.

(Case 2) $\mathrm{Rad}(K) \subsetneq \mathfrak{m}$: we started with the construction and notation as in [10], up to a sign. We shall investigate the function

$$f(z) = -\frac{K(z, z)}{F^2(z)},$$

which is nonnegative on $\mathfrak{m} \setminus \{0\}$. This function is homogeneous and it is reasonable to restrict the definition domain to the indicatrix

$$I_F = \{z \in \mathfrak{m}; F(z) = 1\}.$$

The function $f(z)$ attains its maximum λ_1 at $y_1 \in I_F$. To find the second vector is more delicate. Since the group H is compact and $\mathrm{Rad}(K)$ is an $\mathrm{Ad}(H)$-invariant subspace, there exists an $\mathrm{Ad}(H)$-invariant K-orthogonal complement W of $\mathrm{Rad}(K)$ in $\mathfrak{m}$. Each vector $z \in \mathfrak{m}$ can be uniquely decomposed as $z = z_1 + z_2$, where $z_1 \in \mathrm{Rad}(K)$ and $z_2 \in W$. Denote $k = \dim(\mathrm{Rad}(K))$ and let

$$D_k = \{z_1 \in \mathrm{Rad}(K), F(z_1) < 1\}$$

be the open unit disc in $\mathrm{Rad}(K)$. For each fixed $z_1 \in D_k$, consider the set

$$S_{z_1} = \{z_2 \in W, F(z_1 + z_2) = 1\},$$

which has the topology of a sphere. From now on, if not stated otherwise, $z_1 + z_2$ means $z_1 \in D_k$, $z_2 \in S_{z_1}$ and $z_1 + z_2 \in I_F$. Because $-K > 0$ on W, the function $f(z_1 + z_2)$ is positive for any $z_1 \in D_k$ and $\lim_{z_1 \to \partial D_k} f(z_1 + z_2) = 0$. For fixed z_1 and with definition domain S_{z_1}, $f(z_1 + z_2)$ attains its minimum $\varepsilon(z_1) > 0$ at some $\bar{z}_2(z_1) \in S_{z_1}$. For each $z_1 \in D_k$, we choose one such $\bar{z}_2$ and consider the mapping $\varphi \colon D_k \to I_F$, $z_1 \mapsto z_1 + \bar{z}_2$. The function $f(\varphi(z_1)) = \varepsilon(z_1)$ is smooth on D_k and it attains its maximum λ_2 at $\bar{z}_1$. Here $\bar{z}_1$ can be chosen and the map φ can be defined in a way that there is a neighbourhood $U \subset D_k$ of $\bar{z}_1$ such that the mapping $\varphi|_U$ is smooth. We put $y_2 = \varphi(\bar{z}_1) \in I_F$.

It remains to show that y_1 and y_2 are geodesic vectors. As to y_1, the function

$$\tilde{f}(z) = K(z,z) + \lambda_1 F^2(z)$$

attains its minimum 0 at y_1. For any fixed $w \in \mathfrak{m}$, the function $\hat{f}(t) = \tilde{f}(y_1 + tw)$ attains its minimum 0 at $t = 0$ and hence $\hat{f}'(0) = 0$. Using Formula (1), it follows that

$$-K(y_1, w) = \lambda_1 \cdot g_{y_1}(y_1, w), \qquad \forall w \in \mathfrak{m}$$

and the formula

$$g_{y_1}(y_1, [y_1, z]_{\mathfrak{m}}) = \frac{-1}{\lambda} K(y_1, [y_1, z]_{\mathfrak{m}}) = \frac{-1}{\lambda} K([y_1, y_1], z) = 0, \qquad \forall z \in \mathfrak{m}$$

shows that y_1 is a geodesic vector. As to y_2, we have to modify this approach. The function

$$\tilde{f}(z) = K(z,z) + \lambda_2 F^2(z)$$

attains value 0 at y_2. For fixed $u \in W$, the function $\hat{f}(t) = \tilde{f}(y_2 + tu)$ attains its maximum 0 at $t = 0$ and hence $\hat{f}'(0) = 0$. It follows that

$$-K(y_2, u) = \lambda_2 \cdot g_{y_2}(y_2, u), \qquad \forall u \in W. \tag{2}$$

Now, let $v \in \mathrm{Rad}(K)$ be arbitrary fixed vector. Recall that $y_2 = z_1 + z_2$. Consider the line $z_1 + tv$ in $\mathrm{Rad}(K)$, the curve $c(t) = \varphi(z_1 + tv)$ in I_F and denote by $\bar{v}$ the tangent vector to $c(t)$ at $t = 0$. The function $\hat{f}(t) = \tilde{f}(c(t))$ attains its minimum 0 at $t = 0$ and hence $\hat{f}'(0) = 0$. It follows that

$$-K(y_2, \bar{v}) = \lambda_2 \cdot g_{y_2}(y_2, \bar{v}). \tag{3}$$

Consider a basis $\{u_i\}$ of W, a basis $\{v_j\}$ of $\text{Rad}(K)$ and construct vectors $\bar{v}_j$ as above. It is easy to see that $\{u_i, \bar{v}_j\}$ is a basis of $\mathfrak{m}$ and hence Formulas (2) and (3) for each vector $\bar{v}_j$ imply

$$-K(y_2, w) \;=\; \lambda_2 \cdot g_{y_2}(y_2, w), \qquad \forall w \in \mathfrak{m}.$$

We finish the proof with the formula

$$g_{y_2}(y_2, [y_2, z]_\mathfrak{m}) = \frac{-1}{\lambda_2} K(y_2, [y_2, z]_\mathfrak{m}) = \frac{-1}{\lambda_2} K([y_2, y_2], z) = 0, \qquad \forall z \in \mathfrak{m},$$

which shows that y_2 is a geodesic vector. $\square$

Funding: Research was funded by Internal Grant Agency of Institute of Technology and Business in České Budějovice grant number IGS 8210-009.

Conflicts of Interest: The author declares no conflict of interest. The funders had no role in the design of the study; in the collection, analyses, or interpretation of data; in the writing of the manuscript, or in the decision to publish the results.

References

1. Dušek, Z. Homogeneous geodesics and g.o. manifolds. *Note Mat.* **2018**, *38*, 1–15.
2. Kowalski, O.; Szenthe, J. On the existence of homogeneous geodesics in homogeneous Riemannian manifolds. *Geom. Dedicata* **2000**, *81*, 209–214; Erratum: *Geom. Dedicata* **2001**, *84*, 331–332. [CrossRef]
3. Kowalski, O.; Nikčević, S. and Vlášek, Z. *Homogeneous Geodesics in Homogeneous Riemannian Manifolds—Examples*; Preprint Reihe Mathematik, TU Berlin, No. 665/2000; TU Berlin: Berlin, Germany, 2000.
4. Kowalski, O.; Vlášek, Z. Homogeneous Riemannian manifolds with only one homogeneous geodesic. *Publ. Math. Debrecen* **2003**, *62*, 437–446.
5. Dušek, Z. The existence of homogeneous geodesics in homogeneous pseudo-Riemannian and affine manifolds. *J. Geom. Phys.* **2010**, *60*, 687–689. [CrossRef]
6. Dušek, Z. The existence of light-like homogeneous geodesics in homogeneous Lorentzian manifolds. *Math. Nachr.* **2015**, *288*, 872–876. [CrossRef]
7. Yan, Z.; Deng, S. Existence of homogeneous geodesics on homogeneous Randers spaces. *Houston J. Math.* **2018**, *44*, 481–493.
8. Dušek, Z. The affine approach to homogeneous geodesics in homogeneous Finsler spaces. *Archivum Mathematicum (Brno)* **2018**, *54*, 127–133. [CrossRef]
9. Dušek, Z. The existence of homogeneous geodesics in special homogeneous Finsler spaces. *Matematički Vesnik* **2019**, *71*, 16–22.
10. Yan, Z.; Huang, L. On the existence of homogeneous geodesic in homogeneous Finsler spaces. *J. Geom. Phys.* **2018**, *124*, 264–267. [CrossRef]
11. Dušek, Z. Homogeneous Randers spaces admitting just two homogeneous geodesics. *Archivum Mathematicum (Brno)* **2019**, in press.
12. Shen, Z. *Lectures on Finsler Geometry*; World Scientific: Singapore, 2001.
13. Bao, D.; Chern, S.-S.; Shen, Z. *An Introduction to Riemann-Finsler Geometry*; Springer Science+Business Media: New York, NY, USA, 2000.
14. Deng, S. *Homogeneous Finsler Spaces*; Springer Science+Business Media: New York, NY, USA, 2012.
15. Kowalski, O.; Vanhecke, L. Riemannian manifolds with homogeneous geodesics. *Boll. Un. Math. Ital.* **1991**, *5*, 189–246.
16. Latifi, D. Homogeneous geodesics in homogeneous Finsler spaces. *J. Geom. Phys* **2007**, *57*, 1421–1433. [CrossRef]

symmetry

MDPI

Article

Multivariate Optimal Control with Payoffs Defined by Submanifold Integrals

Andreea Bejenaru *,† and **Constantin Udriste** †

Department of Mathematics and Informatics, University Politehnica of Bucharest, 060042 Bucharest, Romania
* Correspondence: bejenaru.andreea@yahoo.com
† These authors contributed equally to this work.

Received: 7 June 2019; Accepted: 4 July 2019; Published: 8 July 2019

Abstract: This paper adapts the multivariate optimal control theory to a Riemannian setting. In this sense, a coherent correspondence between the key elements of a standard optimal control problem and several basic geometric ingredients is created, with the purpose of generating a geometric version of Pontryagin's maximum principle. More precisely, the local coordinates on a Riemannian manifold play the role of evolution variables ("multitime"), the Riemannian structure, and the corresponding Levi–Civita linear connection become state variables, while the control variables are represented by some objects with the properties of the Riemann curvature tensor field. Moreover, the constraints are provided by the second order partial differential equations describing the dynamics of the Riemannian structure. The shift from formal analysis to optimal Riemannian control takes deeply into account the symmetries (or anti-symmetries) these geometric elements or equations rely on. In addition, various submanifold integral cost functionals are considered as controlled payoffs.

Keywords: maximum principle; optimal control; Einstein manifold; evolution dynamics; cost functional; submanifold integral

MSC: 49J20; 49N05; 49Q10; 53C05; 53C80

1. Introduction

For many centuries, researchers were preoccupied with finding the perfect description for geometric objects (curves, surfaces, and others) with some optimizing features. Therefore, important problems were phrased and solved. Among these, let us recall:

- The Plateau problem concerning the existence of minimal surfaces with isoperimetric constraints;
- The minimal submanifolds as solutions for the volume optimizing problem;
- The harmonic maps resulting from optimizing the energy functional;
- Dirichlet's principle, which identifies the minimizers of the Dirichlet's energy with the solutions of a Poisson equation subject to boundary constraints;
- Fermats's principle which states that the path followed by some ray of light is the one taking the least time;
- Hilbert's isoperimetric problem, stating that the Einstein manifolds are minimizers for the total scalar curvature, with isoperimetric constraints;
- Dieudonne–Rashevsky type problems referring to optimization of multiple integral cost functionals with first order partial differential equations constraints, with applicability in elasticity (the torsion of a prismatic bar), population dynamics (age structure related models), image processing, and others.

Many of these important problems were solved using calculus of variations. Nevertheless, in the last few decades, the optimal control theory has benefited from a consistent development, providing an improvement of the variational techniques and, ultimately, replacing them. Moreover, an important step forward related to optimal control was made by increasing the dimension of the time variable.

Motivated by this mathematical trend, we appreciate as necessary any consistent approach on optimal control theory in geometric framework as it should be suitable for reanalyzing the classical examples, like those presented above, as well as for defining and solving relevant new problems. It is the basic objective of this paper to give answers to the following questions: Is it possible to provide an unitary approach on optimal control which could lead to general tools or formulas for solving all the mentioned problems and possibly others? What are the convenient ways to phrase optimal control problems in the Riemannian context (more precisely, what type of cost functional could be considered)? Which geometric elements will play the key roles of (multi)time, state, and control variables? Which geometric elements interfere in the constraints? What is the geometric significance of the co-state variables?

The main results of our study are Theorem 1 and Corollaries 1–4, containing a formal approach on the Pontryagin's maximum principle (see [1–4]), for multivariate optimal control problems with various types of submanifolds integral type payoffs. Later, in Corollaries 5–9, they are rephrased for a new class of geometric optimal control problems, continuing the ideas from the paper [5]. Not least, Example 2 reconsiders Hilbert's isoperimetric problem in this newly provided setting, while Example 3 provides an additional argument for the utility of this geometric approach. We point out the idea that our Riemannian optimal control is completely distinct from the geometric optimal control described in [6–8], where the role of the evolution variable was the classical one (time variable), while the state and control variables were assumed to be lying on differentiable manifolds.

Our source of inspiration and the research tools cover the following topics:

- Classical optimal control, meaning the original optimal control theory involving a unique time variable, a cost functional including, in general, a running payoff and a terminal payoff, as well as a set of dynamic constraints expressed by ordinary differential equations as well as static constraints expressed generally by inequalities ([9–12]);

- Various statements of the Pontryagin's maximum principle, via a properly defined Hamiltonian ([1–4,13]);

- Multivariate optimal control, initially considered in connection with Dieudonne–Rashevsky problems which involve payoff functionals expressed via multiple integrals and dynamic constraints expressed by first order partial differential equations (see [14–19]);

- Differential geometry under its general aspects, but, more importantly, Riemannian geometry; the most important elements we borrow from Riemannian geometry are the Riemannian metric, the Levi–Civita linear connection, the curvature tensor field, and the equations describing the way they connect (see [20,21]).

A first attempt in the direction of Riemannian optimal control was related to solving two flow-type optimal control problems in the Riemannian setting: The total divergence of a fixed vector field and the total Laplacian (the gradient flux) of a fixed differentiable function. Both times, the cost functional was a multiple-type integral functional (Riemannian extension of Dieudonne–Rashevsky type problems). This paper extends all these ideas by varying the considered type of cost functionals and by considering second order geometric dynamics.

Reaching the above ideas, as well as the ideas developed throughout this paper, was possible after a consistent analysis of multivariate optimal control problems, from different points of views and more extensively than the preliminary approach initiated by Cesari [14] for Dieudonne–Rashevsky problems. For instance, the multivariate optimal control achieved new dimensions by considering other types of cost functionals (stochastic integrals [22], curvilinear-type integrals [23], or mixt payoffs containing both multiple or curvilinear integrals [24]), as well as various types of evolution dynamics (second order partial differential equations, nonholonomic constraints [25]), or different working techniques (multivariate dynamic programming [26], multivariate needle-shaped variations [24,27]). The applicative features of the multivariate Pontryagin's maximum principle were emphasized in [5], where the minimal submanifolds, the harmonic maps, or the Plateau problem were approached under this new light. In addition, multivariate controllability- and observability-related features were studied in [28], while [29] provides a comparison analysis of various types of cost functionals.

In optimal control issues, the variables involved play distinct roles. In this case, the states represent entities with geometric features (Riemannian metric, linear connection, etc.), and the local coordinates of the manifold are variables of evolution. Usually, an object having the properties of the curvature tensor field plays the role of the control element.

The rest of the paper is organized as follows. Section 2 contains a formal overview regarding the multivariate optimal control theory, introducing the specific terminology and establishes the methodology. Section 3 is a review of geometric elements. Section 4 contains the main results derived from applying the technical results from Section 2 to the geometric framework provided in Section 3. Section 5 contains the conclusions and policy implications.

2. Optimal Control Formalism

2.1. Single-Time Case

We start our approach with recalling the standard statement of an optimal control problem, in its most simple form, by namely using a one-dimensional evolution variable. The purpose of this is just to fix the specific terminology and techniques. Later, these elements will be adapted to multi-dimensional evolution variables and ultimately, to geometric objects, by properly identifying the role of each of them.

Formally, an optimal control problem refers to finding:

$$\max_{c(\cdot) \in \mathcal{C}} J(c(\cdot)) = \int_{t_0}^{T} X^0(x, s(x), c(t)) dx + \chi(T, s(T))$$

subject to:

$$\begin{cases} s_0 \in S; \\ \dot{s} = X(x, s(x), c(x)), \ x \in [0, T]. \end{cases}$$

The nature or the meaning of the elements involved in the expressions above are as follows:

- The real number T is called the final time or horizon; t_0 is called initial time. Usually, $x \in [t_0, T]$ represents the time variable, but this comes just from the fact that the optimal control problems which originated this theory used to have temporal evolutions. We prefer to instead call them evolution variables since this terminology is more compatible with the idea of increasing the dimension (we have even avoided to denote it with t);
- $U \subset R^k$ is called the set of control variables. A function $c : [0, T] \to U$ is called the control strategy. Sometimes there are additional requirements concerning the control strategies (for instance, the local integrability condition or static constraints) resulting the set of admissible strategies $\mathcal{C}$;
- $S \subset R^m$ is called the set of state variables. For a given control strategy $c(\cdot)$ and a given initial state $s_0 \in S$, the solution of the evolution equation $s(\cdot) = s(\cdot, s_0, c(\cdot))$ is called the state trajectory;
- $X^0 : [0, T] \times S \times C \to \mathbb{R}$ is called instantaneous performance index. Moreover, $\chi : [0, T] \times S \to \mathbb{R}$ is called the payoff from the final state;
- The functional J on the set of admissible control strategies is called the cost functional or payoff functional.

2.2. Multivariate Case

This section is dedicated to featuring the general aspects of the multivariate optimal control (in a Euclidean setting). The basic ingredients are $N \subset R^n$ with global coordinates $(x^1, ..., x^n)$, $S \subset R^m$ with global coordinates $(s^1, ..., s^m)$, and $U \subset R^k$ having global coordinates $(c^1, ..., c^k)$. Let us denote by D a bounded Lipschitz domain of a p-dimensional submanifold of N, with a $(p-1)$-dimensional oriented boundary ∂D. In particular, when $p = n$ we denote by Ω a bounded Lipschitz domain in N, when $p = n - 1$ we denote by Σ a bounded oriented hyper-surface while, we use C to denote a differentiable curve in N with given endpoints x_i and x_f.

Let $X = (X_i^\alpha) : N \times S \times C \to R^{mn}$ be a C^1 tensor field. For a given control function $c : N \to U$, we define the following completely integrable evolution system:

$$\frac{\partial s^\alpha}{\partial x^i}(t) = X_i^\alpha(x, s(x), c(x)), \ x \in N. \tag{1}$$

The multivariate evolution system in Equation (1) is used as constraint when we want to optimize various integral-type cost functionals.

Problem 1. *p-Dimensional integral cost functional.*

This section reflects the most general expression of multivariate optimal control problems, by considering p-dimensional domains in N and cost functionals defined as integrals on these domains. Denote:

$$\mathcal{I}_\sigma = \begin{cases} \{(i_1 i_2 ... i_{n-\sigma}) \, | \, 1 \leq i_1 < i_2 < ... < i_{n-\sigma} \leq n\}, \ \sigma = \overline{1, n-1}, & p \leq n - 1; \\ \varnothing, & p = n. \end{cases}$$

We define the cost functional:

$$J_D[c(\cdot)] = \int_D \sum_{I \in \mathcal{I}_p} X^I(x, s(x), c(x)) dx_I + \int_{\partial D} \sum_{I \in \mathcal{I}_{p-1}} \chi^I(x, s(x)) dx_I,$$

where, if $I = (i_1 i_2 ... i_{n-p})$, then dx_I is the p-form resulted from the multiple interior product of the n-form dx with the vector fields $\partial_{i_{n-p}}, ..., \partial_{i_1}$.

The corresponding control Hamiltonian $(n-p)$-form has the components

$$H^I(x, s, p, c) = X^I(x, s, c) + p_\alpha^{Is} X_s^\alpha(x, s, c), \ \forall I \in \mathcal{I}_p.$$

In order to keep the expressions as simple as possible, let us introduce the following notations: Given a multi-index $I = (i_1 i_2 ... i_{n-p}) \in \mathcal{I}_p$, let $\partial_I = \partial_{i_1} \wedge \partial_{i_2} \wedge ... \wedge \partial_{i_{n-p}}$, let G denote the induced inner product on the exterior algebra of vector fields, $N_1 \wedge ... \wedge N_{n-p}$ be the cross (wedge) product of the normal distribution on submanifold D, while $\{\eta_1, \eta_2, ..., \eta_{n-p+1}\}$ denotes a normal distribution on ∂D.

Theorem 1. (Multivariate maximum principle for p-dimensional integral cost functional) *Suppose $c^*(\cdot)$ is optimal for $\left(\max\limits_{c(\cdot)} J_D, \text{Equation (1)} \right)$ and $s^*(\cdot)$ is the corresponding optimal n-sheet. Then there exists a costate mapping $(p^*) = (p_\alpha^{*I}) : D \to R^{mn^{n-p}}$, $p_\alpha^{*I} = -p_\alpha^{*\tau(I)}$, $\forall \tau(I)$ a transposition of the multi-index $I \in \mathcal{I}_p$, such that the following equations are satisfied:*

- *State equations:*

$$\frac{\partial s^{*\alpha}}{\partial x^i} = \frac{\partial H^I}{\partial p_\alpha^{Ii}}(x, s^*, p^*, c^*), \ \forall I \in \mathcal{I}_p, \ \forall i = \overline{1, n}, \ \forall \alpha = \overline{1, m}, \ \forall x \in D;$$

- *Adjoint equations:*

$$G\left(\left[\frac{\partial p_\alpha^{*Is}}{\partial x^s} + \frac{\partial H^I}{\partial s^\alpha}(x, s^*, p^*, c^*) \right] \partial_I, N_1 \wedge ... \wedge N_{n-p} \right) = 0, \ \forall \alpha = \overline{1, m}, \ \forall x \in D;$$

- *Optimality conditions*

$$G\left(\frac{\partial H^I}{\partial c^a}(x, s^*, p^*, c^*)\partial_I, N_1 \wedge \ldots \wedge N_{n-p}\right) = 0, \ \forall a = \overline{1, k}, \ \forall x \in D.$$

- *The boundary conditions*

$$G\left(\left[p_\alpha^{*I} - \frac{\partial \chi^I}{\partial s^\alpha}\right]\partial_I, \eta_1 \wedge \ldots \wedge \eta_{n-p+1}\right) = 0, \ \forall \alpha \in \overline{1, m}, \ \forall x \in \partial D.$$

Proof. If $c^*(\cdot)$ is an optimal control, consider a variation $c_\epsilon(\cdot) = c^*(\cdot) + \epsilon v(\cdot)$, $\epsilon \in (-\epsilon_0, \epsilon_0)$. This generates a variational state $s_\epsilon(\cdot)$, with $\left.\frac{ds_\epsilon^\alpha}{d\epsilon}\right|_{\epsilon=0} = \tau^\alpha$ and a cost function:

$$J_\epsilon = J_D[c_\epsilon(\cdot)] = \int_D \sum_{I \in \mathcal{I}_p} X^I(x, s_\epsilon, c_\epsilon)dx_I + \int_{\partial D} \sum_{I \in \mathcal{I}_{p-1}} \chi^I(x, s_\epsilon)dx_I$$

$$= \int_D \sum_{I \in \mathcal{I}_p} \left[H^I(x, s_\epsilon, p, c_\epsilon) - p_\alpha^{Is} X_s^\alpha(x, s_\epsilon, c_\epsilon)\right]dx_I + \int_{\partial D} \sum_{I \in \mathcal{I}_{p-1}} \chi^I(x, s_\epsilon)dx_I$$

$$= \int_D \sum_{I \in \mathcal{I}_p} \left[H^I(x, s_\epsilon, p, c_\epsilon) - p_\alpha^{Is}\frac{\partial s_\epsilon^\alpha}{\partial x^s}\right]dx_I + \int_{\partial D} \sum_{I \in \mathcal{I}_{p-1}} \chi^I(x, s_\epsilon)dx_I$$

$$= \int_D \sum_{I \in \mathcal{I}_p} \left[H^I(x, s_\epsilon, p, c_\epsilon) - \frac{\partial\left(p_\alpha^{Is} s_\epsilon^\alpha\right)}{\partial x^s} + \frac{\partial p_\alpha^{Is}}{\partial x^s}s_\epsilon^\alpha\right]dx_I + \int_{\partial D} \sum_{I \in \mathcal{I}_{p-1}} \chi^I(x, s_\epsilon)dx_I$$

$$= \int_D \sum_{I \in \mathcal{I}_p} \left[H^I(x, s_\epsilon, p, c_\epsilon) + \frac{\partial p_\alpha^{Is}}{\partial x^s}s_\epsilon^\alpha\right]dx_I + \int_{\partial D} \sum_{I \in \mathcal{I}_{p-1}} \left[-p_\alpha^I s^\alpha + \chi^I(x, s_\epsilon)\right]dx_I.$$

Since c^* is a optimal solution, it follows that $\epsilon = 0$ is a critical point for $\epsilon \to J_\epsilon$. That is:

$$0 = \int_D \sum_{I \in \mathcal{I}_p} \left[\left(\frac{\partial H^I}{\partial s^\alpha}(x, s^*, p, c^*) + \frac{\partial p_\alpha^{Is}}{\partial x^s}\right)\tau^\alpha + \frac{\partial H^I}{\partial c^a}(x, s^*, p, c^*)v^a\right]dx_I$$

$$+ \int_{\partial D} \sum_{I \in \mathcal{I}_{p-1}} \left[-p_\alpha^I + \frac{\partial \chi^I}{\partial s^\alpha}(x, s^*)\right]\tau^\alpha dx_I.$$

Choosing the costate tensor p^* as solution for the adjoint partial differential equations system:

$$G\left(\left[\frac{\partial p_\alpha^{Is}}{\partial x^s} + \frac{\partial H^I}{\partial s^\alpha}(x, s^*, p, c^*)\right]\partial_I, N_1 \wedge \ldots \wedge N_{n-p}\right) = 0, \ \forall \alpha = \overline{1, m}, \ \forall x \in D$$

with a boundary condition:

$$G\left(\left[p_\alpha^I - \frac{\partial \chi^I}{\partial s^\alpha}\right]\partial_I, \eta_1 \wedge \ldots \wedge \eta_{n-p+1}\right) = 0, \forall x \in \partial D,$$

we find:

$$\int_D \sum_{I \in \mathcal{I}_p} \frac{\partial H^I}{\partial c^a}(x, s^*, p^*, c^*)v^a dx_I = 0, \ \forall v^a,$$

leading to the optimality conditions:

$$G\left(\frac{\partial H^I}{\partial c^a}(x, s^*, p^*, c^*)\partial_I, N_1 \wedge \ldots \wedge N_{n-p}\right) = 0, \ \forall a = \overline{1, k}, \ \forall x \in D.$$

$\square$

Remark 1. *A better way to phrase the optimality conditions is given by the inequality*

$$\int_D \left[H^I(x, s^*, p^*, c^*) - H^I(x, s^*, p^*, c) \right] dx_I \geq 0, \ \forall c(\cdot) \in \mathcal{C}.$$

A proof leading to this condition is based on needle-shaped control variations and, beside the fact that provides a more general formula, it also allows control variables to reach boundary values (for more details about this technique, please see see [5,24]). Moreover, this expression is preferable when the Hamiltonians are linear with respect to the control variables, i.e., $H^I(x, s, p, c) = \sigma^I(x, s, p) \cdot c + \psi^I(x, s, p)$. If such is the case, two approaches are possible. If $\sigma(x, s(x), p(x)) = 0$ almost everywhere on D, the problem is control-free and the optimal solutions are of a singular-type. Otherwise, the optimal control is a bang-bang, meaning that it switches abruptly between boundary values.

Problem 2. *Multiple integral cost functional.*

This is a particular case of the general one analyzed above, since $\Omega \subset N$ can be considered as a domain of maximal dimension $p = n$. The general expression for a multiple integral cost functional is:

$$J_\Omega[c(\cdot)] = \int_\Omega X(x, s(x), c(x)) \, dx + \int_{\partial\Omega} \chi^I(x, s(x)) \, dx_I.$$

and the corresponding Hamiltonian function (0-form) is:

$$H(x, s, p, c) = X(x, s, c) + p_\alpha^s X_s^\alpha(x, s, c).$$

Then, Theorem 1 reads as in the following Corollary.

Corollary 1. (Multitime maximum principle for multiple integral cost functional) *Suppose $c^*(\cdot)$ is an optimal solution of the control problem $\left(\max\limits_{c(\cdot)} J_\Omega, \text{Equation (1)} \right)$ and $t^*(\cdot)$ is the corresponding optimal state. Then there exists a costate tensor $p^* = (p_\alpha^{*i}) : \Omega \to R^{mn}$ to satisfy:*

- *State equations*

$$\frac{\partial s^{*\alpha}}{\partial x^i} = \frac{\partial H}{\partial p_\alpha^i}(x, s^*, p^*, c^*), \ \forall x \in \Omega, \ \forall \alpha = \overline{1, m}, \ \forall i = \overline{1, n};$$

- *Adjoint equations*

$$\frac{\partial p_\alpha^{*s}}{\partial x^s} = -\frac{\partial H}{\partial s^\alpha}(x, s^*, p^*, c^*), \ \forall x \in \Omega, \ \forall \alpha = \overline{1, m};$$

- *Optimality conditions*

$$\frac{\partial H}{\partial c^a}(x, s^*, p^*, c^*) = 0, \ \forall x \in \Omega, \ \forall a = \overline{1, k};$$

- *Boundary conditions*

$$\left[-p_\alpha^{*I} + \frac{\partial \chi^I}{\partial s^\alpha}(x, s^*) \right] \frac{\partial}{\partial x^I} \in T_x \partial\Omega, \ \forall x \in \partial\Omega, \ \forall \alpha = \overline{1, m}.$$

Problem 3. *Hyper-surface integral cost functional.*

When considering o domain Σ of dimension $p = n - 1$ it results in the following cost functional:

$$J_\Sigma[c(\cdot)] = \int_\Sigma X^I(x, s(x), c(x)) dx_I + \int_{\partial\Sigma} \sum_{1 \leq i < j \leq n} \chi^{ij}(x, s(x)) dx_{ij},$$

where $dx_l = i_{\frac{\partial}{\partial x^l}} dx$ and $dx_{ij} = i_{\frac{\partial}{\partial x^j}} dx_i$.

Similar to previous paragraphs, the multitime maximum principle involves some appropriate Hamiltonian vector field with components:

$$H^l(x,s,p,c) = X^l(x,s,c) + p_\alpha^{ls} X_s^\alpha(x,s,c),$$

and Theorem 1 conducts to the next statement.

Corollary 2. (Multitime maximum principle for hyper-surface integral cost functional) *Suppose $c^*(\cdot)$ is optimal for $\left(\max\limits_{c(\cdot)} J_\Sigma, \text{Equation (1)} \right)$ and $s^*(\cdot)$ is the corresponding optimal n-sheet. Then there exists a co-state mapping $(p^*) = (p_\alpha^{*ij}) : \Sigma \to R^{mn}$, $p_\alpha^{*ij} = -p_\alpha^{*ji}$ to satisfy:*

- *State equations*

$$\frac{\partial s^{*\alpha}}{\partial x^i} = \frac{\partial H^l}{\partial p_\alpha^{li}}(x,s^*,p^*,c^*), \ \forall i = \overline{1,n}, \ \forall \alpha = \overline{1,m}, \ \forall x \in \Sigma;$$

- *Adjoint equations*

$$\left[\frac{\partial p_\alpha^{*ls}}{\partial x^s} + \frac{\partial H^l}{\partial s^\alpha}(x,s^*,p^*,c^*) \right] \frac{\partial}{\partial x^l} \in T_x\Sigma, \ \forall \alpha = \overline{1,m}, \ \forall x \in \Sigma;$$

- *Optimality conditions*

$$\frac{\partial H^l}{\partial c^a}(x,s^*,p^*,c^*) \frac{\partial}{\partial x^l} \in T_x\Sigma, \ \forall a = \overline{1,k}, \ \forall x \in \Sigma;$$

- *Boundary condition*

$$G \left(\sum_{1 \le i < j \le n} \left[-p_\alpha^{*ij} + \frac{\partial \chi^{ij}}{\partial s^\alpha}(x,s^*) \right] \frac{\partial}{\partial x^i} \wedge \frac{\partial}{\partial x^j}, \eta_1 \wedge \eta_2 \right) = 0, \ \forall x \in \partial\Sigma,$$

where G denotes the induced inner product on the exterior algebra of vector fields and $\eta_1 \wedge \eta_2$ is the cross product of the normal distribution $\{\eta_1, \eta_2\}$ on $\partial\Sigma$.

Problem 4. *Curvilinear integral cost functional.*

When the selected domain is a curve C, the corresponding dimension is $p = 1$. The expression of the curvilinear integral cost functional is:

$$J_C[c(\cdot)] = \int_C X_l(x,s(x),c(x))\,dx^l + \chi(x_f,s(x_f)) - \chi(x_i,s(x_i)),$$

where x_i and x_f are the endpoints of C.

The corresponding Hamiltonian is an 1-form with components:

$$H_l(x,s,p,c) = X_l(x,s,c) + p_\alpha X_l^\alpha(x,s,c)$$

leading to the following statement for the maximum principle.

Corollary 3. (Multitime maximum principle for curvilinear integral cost functional) *Suppose $c^*(\cdot)$ is an optimal solution of the control problem $\left(\max\limits_{c(\cdot)} J_C, \text{Equation (1)} \right)$ and $s^*(\cdot)$ is the corresponding optimal state. Then there exists a costate mapping $p^* = (p_\alpha^*) : C \to R^m$ to satisfy:*

- State equations

$$\frac{\partial s^{*\alpha}}{\partial x^l} = \frac{\partial H_l}{\partial p_\alpha}(x,s^*,p^*,c^*), \ \forall x \in C, \ \forall \alpha = \overline{1,m}, \ \forall l = \overline{1,n};$$

- Adjoint equations

$$\delta^{ls} \left[\frac{\partial p_\alpha^*}{\partial x^l} + \frac{\partial H_l}{\partial s^\alpha}(x,s^*,p^*,c^*) \right] \frac{\partial}{\partial x^s} \in T^\perp C, \ \forall x \in C, \ \forall \alpha = \overline{1,m};$$

- Optimality conditions

$$\delta^{ls} \frac{\partial H_l}{\partial u^a}(x,s^*,p^*,c^*) \frac{\partial}{\partial x^s} \in T^\perp C, \ \forall x \in C, \ \forall a = \overline{1,k};$$

- Terminal conditions

$$\begin{cases} p_\alpha^*(x_f) = \dfrac{\partial \chi}{\partial s^\alpha}(x_f, s^*(x_f)); \\ p_\alpha^*(x_i) = \dfrac{\partial \chi}{\partial s^\alpha}(x_i, s^*(x_i)), \ \forall \alpha \in \overline{1,m}. \end{cases}$$

Problem 5. *Evolution equations with symmetries.*

The previous sections phrased optimal control conditions for the evolution system in Equation (1) and for different types of integral costs. The section instead aims to describe the optimal control behavior, when dealing with an evolution system supporting some sort of symmetries. Assume that the dimension of the considered domain D is $p \geq 2$. We define:

1. A symmetric-type evolution system:

$$\frac{\partial s_j^\alpha}{\partial x^i}(x) + \frac{\partial s_i^\alpha}{\partial x^j}(x) = X_{ij}^\alpha(x,s(x),c(x)) + X_{ji}^\alpha(x,s(x),c(x)). \tag{2}$$

2. An ntisymmetric-type evolution system:

$$\frac{\partial s_j^\alpha}{\partial x^i}(x) - \frac{\partial s_i^\alpha}{\partial x^j}(x) = X_{ij}^\alpha(x,s(x),c(x)) - X_{ji}^\alpha(x,s(x),c(x)). \tag{3}$$

The multivariate maximum principles (necessary conditions) corresponding to the optimal control problems $\left(\max\limits_{c(\cdot)} J_D, Equation\ (2) \right)$ and $\left(\max\limits_{c(\cdot)} J_D, Equation\ (3) \right)$ connects the existence of an optimal control c^* to co-state mappings $p^* = (p_\alpha^{*I})_{I \in \mathcal{I}_{p-2}}$, with some symmetry particularities:

(p1) in the case of symmetric-type evolution system, $p^I = -p^{\tau(I)}$ for each transposition of the multi-index $I \in \mathcal{I}_{p-2}$, except the transposition τ_0 of the last two elements of the multi-index, for which $p^I = p^{\tau_0(I)}$;

(p2) in the case of antisymmetric-type evolution system, $p^I = -p^{\tau(I)}$ for each transposition of the multi-index $I \in \mathcal{I}_{p-2}$, with no exceptions.

These costate mappings allow the definition of the Hamiltonian $(n-p)$-form of components:

$$H^I(x,s,p,c) = X^I(x,s,c) + p_\alpha^{Iij} X_{ij}^\alpha(x,s,c), \ \forall I \in \mathcal{I}_p.$$

Using their symmetries, similar arguments as in the proof of Theorem 1 lead to the outcome stated below.

Corollary 4. (multitime maximum principle for symmetric/antisymmetric evolution equations)
Suppose $c^*(\cdot)$ *is optimal for* $\left(\max\limits_{c(\cdot)} J_D, Equation\ (2) \right)$ *or* $\left(\max\limits_{c(\cdot)} J_D, Equation\ (3) \right)$ *and that* $t^*(\cdot)$ *is the*

corresponding optimal n-sheet. Then there exists a co-state mapping p* with properties (p_1) and (p_2), respectively, to satisfy:

- State equations:

$$\frac{\partial s^{*\alpha}_j}{\partial x^i} \pm \frac{\partial s^{*\alpha}_i}{\partial x^j} = \left[\frac{\partial H^I}{\partial p^{Iij}_\alpha} \pm \frac{\partial H^I}{\partial p^{Iji}_\alpha} \right] (x, s^*, p^*, c^*),$$

$$\forall x \in D, \ \forall I \in \mathcal{I}_p, \ \forall i, j = \overline{1, n}, \ \forall \alpha = \overline{1, m};$$

- Adjoint equations:

$$G\left(\left[\frac{\partial p^{*Isi}_\alpha}{\partial x^s} + \frac{\partial H^I}{\partial s^\alpha_i}(x, s^*, p^*, c^*) \right] \partial_I, N_1 \wedge \dots \wedge N_{n-p} \right) = 0,$$

$$\forall x \in \Omega, \ \forall i = \overline{1, n}, \ \forall \alpha = \overline{1, m};$$

- Optimality conditions:

$$G\left(\frac{\partial H^I}{\partial u^a}(x, s^*, p^*, c^*) dx_I, N_1 \wedge \dots \wedge N_{n-p} \right) = 0, \ \forall x \in D, \ \forall a = \overline{1, k};$$

- Boundary conditions:

$$G\left(\left[p^{*Il}_\alpha - \frac{\partial \chi^l}{\partial s^\alpha_l}(x, s^*) \right] \partial_I, \eta_1 \wedge \dots \wedge \eta_{n-p+1} \right) = 0, \ \forall x \in \partial D, \ \forall \alpha = \overline{1, m}, \ \forall l = \overline{1, n}.$$

3. Basics on Riemannian Geometry

Let (M, g) be a Riemannian manifold and $(x^1, \dots, x^n)$ be local coordinates on M. A basic result in Riemannian geometry ([20,21]) states the existence of the Levi–Civita connection, i.e., the unique torsion-free ($\nabla_X Y - \nabla_Y X = [X, Y]$) and metric compatible ($\nabla g = 0$) linear connection ∇ associated to g.

In coordinates, the Levi–Civita connection can be described using the Christoffel symbols $\Gamma = \left(\Gamma^k_{ij} \right)$. The torsion free condition is then equivalent to the symmetry property $\Gamma^k_{ij} = \Gamma^k_{ji}$, while the compatibility with the metric is given by the following partial differential equations:

$$\frac{\partial g_{ij}}{\partial x^k}(x) = g_{ps}(x) \left[\delta^p_i \Gamma^s_{jk}(x) + \delta^p_j \Gamma^s_{ik}(x) \right], \ i, j, k = 1, \dots, n, \tag{4}$$

or, equivalent,

$$\frac{\partial g^{ij}}{\partial x^k}(x) = -g^{ps}(x) \left[\delta^i_p \Gamma^j_{sk}(x) + \delta^j_p \Gamma^i_{sk}(x) \right], \ i, j, k = 1, \dots, n, \tag{5}$$

where $g^{-1} = (g^{ij})$ is the dual metric tensor field, i.e., $g^{is} g_{sj} = \delta^i_j, \ \forall i, j = 1, \dots, n$.

Moreover, a second order covariant differentiation of the Riemannian structure g generates the Riemann curvature $(1, 3)$-tensor field:

$$R(X, Y)Z = \nabla_X \nabla_Y Z - \nabla_Y \nabla_X Z - \nabla_{[X, Y]} Z,$$

which, in terms of local coordinates $R = (R^l_{ijk})$, is defined by:

$$R^l_{kij} = \frac{\partial \Gamma^l_{kj}}{\partial x^i} - \frac{\partial \Gamma^l_{ki}}{\partial x^j} + \Gamma^s_{kj} \Gamma^l_{si} - \Gamma^s_{ki} \Gamma^l_{sj}, \ i, j, k, l = 1, \dots, n,$$

or, equivalent:

$$\frac{\partial \Gamma^l_{kj}}{\partial x^i} - \frac{\partial \Gamma^l_{ki}}{\partial x^j} = R^l_{kij} - \Gamma^s_{kj}\Gamma^l_{si} + \Gamma^s_{ki}\Gamma^l_{sj}, \ i,j,k,l = 1,\dots,n. \tag{6}$$

Lowering the index via the metric g, allows the introduction of the Riemann curvature $(0,4)$-type tensor field $R_{ijkl} = g_{is}R^s_{jkl}$, having the symmetry properties:

$$R_{ijkl} = R_{klij}; \ R_{ijkl} = -R_{jikl} \tag{7}$$

and satisfying the Bianchi identities:

$$R_{ijkl} + R_{iklj} + R_{iljk} = 0; \ R_{ijkl,r} + R_{ijlr,k} + R_{ijrk,l} = 0, \tag{8}$$

where a comma denotes the covariant derivative. We introduce the set of *curvature like tensor fields*:

$$\mathcal{CT}^0_4 = \{T_{ijkl} \,|\, \text{with the properties from relations (7), (8)}\}.$$

In the following, we shall switch the order of the geometric ingredients. Given a $(0,4)$-tensor field $R = (R_{ijkl})$ in $\mathcal{CT}^0_4$, we ask ourselves whether there exist a linear connection Γ and a Riemannian structure g on M satisfying Equations (4) and (6), respectively Equations (5) and (6). More precisely, adding initial conditions:

$$g_{ij}(x_0) = \eta_{ij}, \ \Gamma^k_{ij}(x_0) = \gamma^k_{ij}(x_0),$$

we consider the relations in Equations (4) and (6) and Equations (5) and (6) as *controlled evolution laws* and we shall call them *second order metric compatibility evolution system*.

Hereafter, the metric tensor $g = (g_{ij})$ and the linear connection $\Gamma = (\Gamma^k_{ij})$ will denote *symmetric state objects*, the local coordinates $x = (x^1,\dots,x^n)$ will play the role of the *evolution variables*, and the tensor field $R = (R_{ijkl})$ will denote a *control object with symmetries*.

The partial differential equations system provided by Equations (4) and (6) has solutions if and only if the complete integrability conditions:

$$\begin{cases} \frac{\partial}{\partial x^l} \left\{ g_{ps} \left[\delta^p_i \Gamma^s_{jk} + \delta^p_j \Gamma^s_{ik} \right] \right\} = \frac{\partial}{\partial x^k} \left\{ g_{ps} \left[\delta^p_i \Gamma^s_{jl} + \delta^p_j \Gamma^s_{il} \right] \right\}; \\ 0 = \frac{\partial}{\partial x^p} \left(R^l_{kij} - \Gamma^s_{kj}\Gamma^l_{si} + \Gamma^s_{ki}\Gamma^l_{sj} \right) + \frac{\partial}{\partial x^i} \left(R^l_{kjp} - \Gamma^s_{kp}\Gamma^l_{sj} + \Gamma^s_{kj}\Gamma^l_{sp} \right) \\ + \frac{\partial}{\partial x^j} \left(R^l_{kpi} - \Gamma^s_{ki}\Gamma^l_{sp} + \Gamma^s_{kp}\Gamma^l_{si} \right) \end{cases}$$

are satisfied. Explicitly, this means $R_{ijkl} = -R_{jikl}$ and $R_{ijkl,r} + R_{ijlr,k} + R_{ijrk,l} = 0$. These relations are among the properties of $R = (R_{ijkl})$ since we have assumed $R = (R_{ijkl})$ to be described by the conditions in Equations (7) and (8).

4. Riemannian Optimal Control

In order to motivate our further approach, we provide the following example from [5], which proves that some problems turn out to be very interesting optimal control issues, by properly stating them and by properly assigning roles for the involved variables.

Example 1. *If D is a compact set of $\mathbb{R}^m = (t^1,\dots,t^m)$, with a piecewise smooth $(m-1)$-dimensional boundary ∂D, then its volume can be expressed as follows:*

$$\mathcal{V}(D) = \int_D dt = \frac{1}{m}\int_{\partial D} \delta_{\alpha\beta} t^\alpha N^\beta d\sigma,$$

where N denotes the exterior unit normal vector field on the boundary. On the other hand, by taking a parametrization of ∂D, having the parameters' domain $U \subset \mathbb{R}^{m-1} = \{\eta_1,\dots,\eta_{m-1}\}$, the area of the boundary surface is:

$$\mathcal{A}(\partial D) = \int_{\partial D} d\sigma = \int_U \sqrt{\delta_{\alpha\beta}\mathcal{N}^\alpha\mathcal{N}^\beta}d\eta,$$

where $\mathcal{N}$ stands for the exterior normal vector field, hence $d\sigma = ||\mathcal{N}||d\eta$.

Let us show that of all solids having a given surface area, the sphere being the one that has the greatest volume. To prove this statement, we take the normal vector field $\mathcal{N}$ as a control and we formulate the multivariate optimal control problem with (static) isoperimetric constraint:

$$\max_{\mathcal{N}} \int_{\partial D} \delta_{\alpha\beta}t^\alpha\mathcal{N}^\beta d\eta \text{ subject to } \int_U \sqrt{\delta_{\alpha\beta}\mathcal{N}^\alpha\mathcal{N}^\beta}d\eta = \text{const.}.$$

The corresponding Hamiltonian is:

$$H(t, p, \mathcal{N}) = \delta_{\alpha\beta}t^\alpha\mathcal{N}^\beta + p\sqrt{\delta_{\alpha\beta}\mathcal{N}^\alpha\mathcal{N}^\beta}, \ p = \text{const.}$$

and the optimality conditions lead to:

$$0 = \frac{\partial H}{\partial N} = t - pN \text{ on } \partial D,$$

which, knowing that $||N|| = 1$, describes the boundary of D as being the solution for $||t||^2 = p^2$. Hence D is precisely the ball of radius p.

If (M, g) is a n-dimensional Riemannian manifold, let $x = (x^1, ..., x^n)$ denote the local coordinates relative to a fixed local map (V, h). We use the same notations as in the formal case: Ω is a bounded Lipschitz domain of M, with oriented boundary $\partial\Omega$, Σ is a bounded oriented hyper-surface, while C denotes a differentiable curve on M with given endpoints x_i and x_f.

We shall further consider several types of cost functionals.

I. Curvature related functionals

1. *Multiple integral-type functional:*

$$J_\Omega[R(\cdot)] = \int_\Omega X(x, g(x), \Gamma(x), R(x))dx + \int_{\partial\Omega} \chi^i(x, g(x), \Gamma(x))dx_i,$$

where $dx = dx^1 \wedge ... \wedge dx^n$ denotes the canonical differential n-form on M and $dx_l = i_{\frac{\partial}{\partial x^l}} dx$, i_X denoting the interior product of a differential form with respect to a vector field X.

2. *Hyper-surface integral-type functional:*

$$J_\Sigma[R(\cdot)] = \int_\Sigma X^l(x, g(x), \Gamma(x), R(x))dx_l + \int_{\partial\Sigma} \sum_{1 \le i < j \le n} \chi^{ij}(x, g(x), \Gamma(x))dx_{ij},$$

where $dx_{ij} = i_{\frac{\partial}{\partial x^j}} dx_i$.

3. *Path independent curvilinear integral-type cost:*

$$J_C[R(\cdot)] = \int_C X_l(x, g(x), \Gamma(x), R(x))dx^l + \chi(x_f, g(x_f), \Gamma(x_f)) - \chi(x_i, g(x_i), \Gamma(x_i)).$$

II. Connection related functionals

4. *I. Multiple integral-type functional:*

$$J_\Omega[\Gamma(\cdot)] = \int_\Omega X(x, g(x), \Gamma(x))dx + \int_{\partial\Omega} \chi^i(x, g(x))dx_i.$$

5. *Hyper-surface integral-type functional:*

$$J_{\Sigma}[\Gamma(\cdot)] = \int_{\Sigma} X^{l}(x, g(x), \Gamma(x)) dx_{l} + \int_{\partial\Sigma} \sum_{1 \leq i < j \leq n} \chi^{ij}(x, g(x)) dx_{ij}.$$

6. *Path independent curvilinear integral-type cost:*

$$J_{C}[\Gamma(\cdot)] = \int_{C} X_{l}(x, g(x), \Gamma(x)) dx^{l} + \chi(x_{f}, g(x_{f})) - \chi(x_{i}, g(x_{i})).$$

Definition 1. *The problem of maximizing (minimizing) one of the cost functionals* $(J_{\Omega}) - (J_{C})$, *subject to one of the metric evolution systems given by Equations* (4) *and* (6) *or Equations* (5) *and* (6) *is called the Riemannian optimal control problem.*

All the outcomes resulted in connection with the functionals above are in fact the expressions from Corollaries 1–3, for the particular choice of the state variables $s = (g, \Gamma)$ and control variables $c = R$ (or $s = g$ and $c = \Gamma$ if the curvature tensor is not involved at all). Since the main ingredients of this Riemannian optimal control problem (the state variables, the control variables, and evolution constraints) have some sort of symmetries, we shall derive adapted multitime maximum principles, based on co-state variables with symmetries as in Corollary 4. In the following, we list these outcomes, together with the Hamiltonians they rely on.

4.1. Riemannian Control with Multiple Integral Cost Functional

Problem 6. *Optimize* $J_{\Omega}[R(\cdot)]$ *subject to Equations* (5) *and* (6).

For that, let us consider Lagrange multipliers of type $p_{ij}^{k} = p_{ji}^{k}$ and $q_{s}^{k\,ij} = -q_{s}^{k\,ji}$ and the control Hamiltonian:

$$H(x, g, \Gamma, R, p, q) = X(x, g, \Gamma, R) - g^{is}\Gamma_{sk}^{j}p_{ij}^{k} + q_{s}^{k\,ij}\left(\frac{1}{2}R_{kij}^{s} - \Gamma_{kj}^{p}\Gamma_{pi}^{s}\right).$$

Corollary 5. *Suppose the tensor field* $R^{*}(\cdot)$ *is an optimal solution for* $\max\limits_{R(\cdot)} J_{\Omega}[R(\cdot)]$, *constraint by the evolution laws in Equations* (5) *and* (6) *and that* $g^{*}(\cdot)$ *and* $\Gamma^{*}(\cdot)$ *are the corresponding optimal Riemannian structure and the optimal linear connection, respectively. Then there exist the dual objects* $p^{*} = (p_{ij}^{*k} = p_{ji}^{*k})$ *and* $q^{*} = (q_{s}^{*k\,ij} = -q_{s}^{*k\,ji})$ *satisfying:*

-   *The state equations:*

$$\begin{cases} \dfrac{\partial g^{*ij}}{\partial x^{k}} = \dfrac{\partial H}{\partial p_{ij}^{k}} + \dfrac{\partial H}{\partial p_{ji}^{k}}, \\[2mm] \dfrac{\partial \Gamma_{kj}^{s}}{\partial x^{i}} - \dfrac{\partial \Gamma_{ki}^{s}}{\partial x^{j}} = \dfrac{\partial H}{\partial q_{s}^{k\,ij}} - \dfrac{\partial H}{\partial q_{s}^{k\,ji}}; \end{cases}$$

-   *The adjoint equations:*

$$\begin{cases} \dfrac{\partial p_{ij}^{*k}}{\partial x^{k}} + \left(\dfrac{\partial H}{\partial g^{ij}} + \dfrac{\partial H}{\partial g^{ji}}\right) = 0, \\[2mm] \dfrac{\partial q_{s}^{*i\,kj}}{\partial x^{k}} + \dfrac{\partial q_{s}^{*j\,ki}}{\partial x^{k}} + \left(\dfrac{\partial H}{\partial \Gamma_{ij}^{s}} + \dfrac{\partial H}{\partial \Gamma_{ji}^{s}}\right) = 0; \end{cases}$$

-   *The optimality conditions:*

$$\dfrac{\partial H}{\partial R_{kij}^{s}} - \dfrac{\partial H}{\partial R_{kji}^{s}} = 0;$$

- The boundary conditions:

$$\begin{cases} p_{ij}^{*k}|_{\partial\Omega} = \left[\dfrac{\partial \chi^k}{\partial g^{ij}} + \dfrac{\partial \chi^k}{\partial g^{ji}} \right]_{\partial\Omega}, \\[2mm] \left[q_s^{*ikj} + q_s^{*jki} \right]_{\partial\Omega} = \left[\dfrac{\partial \chi^k}{\partial \Gamma_{ij}^s} + \dfrac{\partial \chi^k}{\partial \Gamma_{ji}^s} \right]_{\partial\Omega}. \end{cases}$$

Remark 2. *If* (g^*, Γ^*, R^*) *is an optimal solution with corresponding dual objects* (p^*, q^*) *and*

$$H_j^{*i} = H_j^i(g^*, \Gamma^*, R^*, p^*, q^*) = X(g^*, \Gamma^*, R^*)\delta_j^i - g^{*ks}\Gamma_{sj}^{*l}p_{kl}^{*i} + q_s^{*kil}\frac{\partial \Gamma_{kl}^{*s}}{\partial x^j}$$

is an autonomous anti-trace Hamiltonian, then the following conservation law is satisfied

$$D_i H_j^{*i} = 0, \ \forall j = \overline{1, n}.$$

Example 2. *(Hilbert's isoperimetric problem) Consider the functional* $I[R(\cdot)] = \rho(\Omega)$, *where* $\rho(\Omega) = \displaystyle\int_\Omega \rho dv$
denotes the total scalar curvature. Therefore, we try to minimize $I[R(\cdot)] = \int_\Omega g^{ij} R_{ikj}^k \sqrt{g} dx$, *subject to the controlled evolution system defined by Equations* (5) *and* (6) *and to the isoperimetric constraint* $vol(\Omega) = C$.
We start by introducing a Lagrangian functional:

$$J_\Omega[R(\cdot)] = \rho(\Omega) - \lambda vol(\Omega) = \int_\Omega \left[g^{ij} R_{ikj}^k \sqrt{g} - \lambda \sqrt{g} \right] dx.$$

We may identify $X(x, g, \Gamma, R) = g^{ij} R_{ikj}^k \sqrt{g} - \lambda \sqrt{g}$ *and* $\chi^k(x, g, \Gamma) = 0$. *The corresponding Hamiltonian density is*

$$H(x, g, \Gamma, R, p, q) = g^{ij} R_{ikj}^k \sqrt{g} - \lambda \sqrt{g} - g^{is}\Gamma_{sk}^j p_{ij}^k + q_l^{kij} \left(\frac{1}{2} R_{kij}^l - \Gamma_{kj}^s \Gamma_{si}^l \right).$$

Denoting $\sigma_l^{kij}(g, \Gamma, p, q) = \dfrac{1}{2} \left(q_l^{kij} + g^{kj}\delta_l^i \sqrt{g} - g^{ki}\delta_l^j \sqrt{g} \right)$ *and*
$\psi(g, \Gamma, p, q) = -\lambda\sqrt{g} - g^{is}\Gamma_{sk}^j p_{ij}^k - q_l^{kij}\Gamma_{kj}^s \Gamma_{si}^l$, *we may rewrite the autonomous Hamiltonian*

$$H(g, \Gamma, R, p, q) = \sigma_l^{kij}(g, \Gamma, p, q) R_{kij}^l + \psi(g, \Gamma, p, q),$$

which is linear with respect to the control variables. For bang-bang optimal control, we impose $||R(\cdot)|| \le M$, *where the norm is the Riemannian one. To judge in the sense of singular optimal control, we need* $\sigma(x) \equiv 0$, $x \in \Omega_1 \subset \Omega$. *Therefore, the optimal solutions may exhibit both bang-bang and singular sub-sheets as described in Remark* 1.
 Let us search for singular solutions (see [9]*), that is* (i) $\sigma(x) \equiv 0$ *and* (ii) *the conservation law for the autonomous anti-trace Hamiltonian is satisfied.*
 The first condition, combined with the antisymmetry property of q, *provides:*

$$q_l^{kij} = \left[g^{ki}\delta_l^j - g^{kj}\delta_l^i \right] \sqrt{g}.$$

In addition to this, the singular solution also satisfies the adjoint partial differential equations system:

$$\frac{\partial p_{ij}^k}{\partial x^k} = p_{is}^k \Gamma_{jk}^s + p_{js}^k \Gamma_{ik}^s + \left[-2R_{ikj}^k + (\rho - \lambda)g_{ij} \right] \sqrt{g}$$

and:

$$\frac{\partial q_l^{ikj}}{\partial x^k} + \frac{\partial q_l^{jki}}{\partial x^k} = p_{sk}^j g^{si} + p_{sk}^i g^{sj} + \Gamma_{sk}^i q_l^{sjk} + \Gamma_{sl}^k q_k^{jsi} + \Gamma_{sk}^j q_l^{sik} + \Gamma_{sl}^k q_k^{isj}.$$

Replacing q in the latter leads to $p^j_{sk}g^{si} + p^i_{sk}g^{sj} = 0$, *with the solution*

$$p^k_{ij} = 0.$$

Finally, by substituting p in the first adjoint set of equations, we obtain $R^k_{ikj} = \dfrac{\rho - \lambda}{2}g_{ij}$, *that is the Einstein Equation in vacuum* $Ric_{ij} = \dfrac{\lambda}{n-2}g_{ij}$.

Moreover, the anti-trace autonomous Hamiltonian is:

$$H^{*i}_j = \left[(\rho - \lambda)\delta^i_j + g^{*ki}\frac{\partial \Gamma^{*s}_{ks}}{\partial x^j} - g^{*ks}\frac{\partial \Gamma^{*i}_{ks}}{\partial x^j}\right]\sqrt{g^*}$$

and the conservation law $D_i H^{*i}_j = 0$ *is satisfied by the Einstein structure, therefore, the Einstein manifolds are singular critical points for the total scalar curvature functional with isoperimetric constraints.*

Problem 7. *Optimize* $J_\Omega[\Gamma(\cdot)]$ *subject to Equation* (5).

The corresponding Hamiltonian has a simplified expression:

$$H(x, g, \Gamma, p, q) = X(x, g, \Gamma) - g^{is}\Gamma^j_{sk}p^k_{ij}$$

and the multitime maximum principle is described by the following Corollary.

Corollary 6. *Suppose the linear connection* $\Gamma^*(\cdot)$ *is an optimal solution for* $\left(\max\limits_{\Gamma(\cdot)} J_\Omega(\Gamma(\cdot), \text{Equation } (5)\right)$ *and that* $g^*(\cdot)$ *is the corresponding optimal Riemannian structure. Then there exist a dual object* $p^* = (p^{*k}_{ij} = p^{*k}_{ji})$ *satisfying:*

- *The state equations:*

$$\frac{\partial g^{*ij}}{\partial x^k} = \frac{\partial H}{\partial p^k_{ij}} + \frac{\partial H}{\partial p^k_{ji}};$$

- *The adjoint equations:*

$$\frac{\partial p^{*k}_{ij}}{\partial x^k} + \left(\frac{\partial H}{\partial g^{ij}} + \frac{\partial H}{\partial g^{ji}}\right) = 0;$$

- *The optimality conditions:*

$$\frac{\partial H}{\partial \Gamma^k_{ij}} + \frac{\partial H}{\partial \Gamma^k_{ji}} = 0;$$

- *The boundary conditions:*

$$p^{*k}_{ij}|_{\partial\Omega} = \left[\frac{\partial \chi^k}{\partial g^{ij}} + \frac{\partial \chi^k}{\partial g^{ji}}\right]_{\partial\Omega}.$$

Example 3. *Consider the least squares Lagrangian-type cost functional:*

$$J[\Gamma] = \frac{1}{2}\int_\Omega g^{ij}\Gamma^k_{is}\Gamma^s_{jk}dx,$$

which measures the mean square deviation tensor $\Gamma - \Gamma^0$, *where* Γ *is a linear connection and* $\Gamma^0 = 0$ *is the Euclidean linear connection. The corresponding Hamiltonian density is:*

$$H = \frac{1}{2}g^{ij}\Gamma^k_{is}\Gamma^s_{jk} - g^{is}\Gamma^j_{sk}p^k_{ij}$$

and, according to Corollary 6, we have:

- *The optimality conditions:* $g^{is}\Gamma^j_{sk} + g^{js}\Gamma^i_{sk} - g^{is}p^j_{sk} - g^{js}p^i_{sk} = 0$, *leading to the general solution*

$$p^k_{ij} = \Gamma^k_{ij} - \gamma^k_{ij}, \text{ or, invariant } p = \Gamma - T,$$

  where $T = (\gamma^k_{ij})$ is a (1,2) symmetric tensor field, satisfying the anti-symmetry condition $g^{is}\gamma^j_{sk} + g^{js}\gamma^i_{sk} = 0$;
- *The boundary conditions $p^k_{ij}|_{\partial\Omega} = 0$, which, by substituting p, lead to*

$$\gamma^k_{ij}|_{\partial\Omega} = \Gamma^k_{ij}|_{\partial\Omega};$$

- *The adjoint equations:*

$$\frac{\partial p^k_{ij}}{\partial x^k} = -\Gamma^k_{is}\Gamma^s_{jk} + \Gamma^k_{is}p^s_{jk} + \Gamma^k_{js}p^s_{ik},$$

  rewritten, after substituting p: $\dfrac{\partial\Gamma^k_{ij}}{\partial x^k} - \dfrac{\partial\gamma^k_{ij}}{\partial x^k} = \Gamma^k_{is}\Gamma^s_{jk} - \gamma^k_{is}\Gamma^s_{jk} - \gamma^k_{js}\Gamma^s_{ik}, \text{ or, } \dfrac{\partial\Gamma^k_{ij}}{\partial x^k} - \Gamma^k_{is}\Gamma^s_{jk} = (\text{Div }T)_{ij}, \text{ or,}$
  even better

$$\text{Ric}_{ij} + \nabla_{\partial i}\left(\frac{\partial\ln\sqrt{g}}{\partial x^j}\right) = (\text{Div }T)_{ij}.$$

In particular, by taking $\gamma = 0$, it follows that manifolds satisfying

$$\text{Ric} = -\nabla\, d\ln\sqrt{g}$$

are critical points for the functional

$$J[\Gamma(\cdot)] = \frac{1}{2}\int_\Omega g^{ij}\Gamma^k_{is}\Gamma^s_{jk}dx.$$

4.2. Riemannian Control with Hypersurface Integral-Type Cost Functional

Problem 8. *Optimize $J_\Sigma[R(\cdot)]$ subject to Equations (5) and (6).*

Let us consider Lagrange multipliers of type $p^{lk}_{ij} = p^{lk}_{ji} = -p^{kl}_{ij}$ and $q^{lkij}_s = -q^{lkji}_s = -q^{klij}_s$, and the control Hamiltonian vector field:

$$H^l(x, g, \Gamma, p) = X^l(x, g, \Gamma, R) - g^{is}\Gamma^j_{sk}p^{lk}_{ij} + q^{lkij}_s\left(\frac{1}{2}R^s_{kij} - \Gamma^p_{kj}\Gamma^s_{pi}\right).$$

Corollary 7. *Suppose the tensor field $R^*(\cdot)$ is an optimal solution for $\left(\max\limits_{R(\cdot)} J_\Sigma[R(\cdot)], \text{Equations}(5) \text{and}(6)\right)$ and that $g^*(\cdot)$ and $\Gamma^*(\cdot)$ are the corresponding optimal Riemannian structure and the optimal linear connection, respectively. Then there exist the dual objects $p^* = (p^{*lk}_{ij} = p^{*lk}_{ji} = -p^{*kl}_{ij})$ and $q^* = (q^{*lkij}_s = -q^{*lkji}_s = -q^{*klij}_s)$ satisfying:*

- *The state equations:*

$$\begin{cases} \dfrac{\partial g^{*ij}}{\partial x^k} = \dfrac{\partial H^l}{\partial p^{lk}_{ij}} + \dfrac{\partial H^l}{\partial p^{lk}_{ji}}, \ \forall l = \overline{1, n}, \\[3mm] \dfrac{\partial\Gamma^s_{kj}}{\partial x^i} - \dfrac{\partial\Gamma^s_{ki}}{\partial x^j} = \dfrac{\partial H^l}{\partial q^{lkij}_s} - \dfrac{\partial H^l}{\partial q^{lkji}_s}; \end{cases}$$

- *The adjoint equations:*

$$
\begin{cases}
\left[\dfrac{\partial p^{*lk}_{ij}}{\partial x^k} + \left(\dfrac{\partial H^l}{\partial g^{ij}} + \dfrac{\partial H^l}{\partial g^{ji}}\right)\right]\dfrac{\partial}{\partial x^l} \in T_x\Sigma, \\[4mm]
\left[\left(\dfrac{\partial q^{*li\,kj}_s}{\partial x^k} + \dfrac{\partial q^{*lj\,ki}_s}{\partial x^k}\right) + \left(\dfrac{\partial H^l}{\partial \Gamma^s_{ij}} + \dfrac{\partial H^l}{\partial \Gamma^s_{ji}}\right)\right]\dfrac{\partial}{\partial x^l} \in T_x\Sigma;
\end{cases}
$$

- *The optimality conditions:*

$$
\left[\dfrac{\partial H^l}{\partial R^s_{kij}} - \dfrac{\partial H^l}{\partial R^s_{kji}}\right]\dfrac{\partial}{\partial x^l} \in T_x\Sigma;
$$

- *The boundary conditions:*

$$
\begin{cases}
G\left(\left[p^{*lk}_{ij} - \left(\dfrac{\partial \chi^{lk}}{\partial g^{ij}} + \dfrac{\partial \chi^{lk}}{\partial g^{ji}}\right)\right]\partial_{lk}, \eta_1 \wedge \eta_2\right) = 0, \\[4mm]
G\left(\left[\left(q^{*li\,kj}_s + q^{*lj\,ki}_s\right) - \left(\dfrac{\partial \chi^{lk}}{\partial \Gamma^s_{ij}} + \dfrac{\partial \chi^{lk}}{\partial \Gamma^s_{ji}}\right)\right]\partial_{lk}, \eta_1 \wedge \eta_2\right) = 0,
\end{cases}
$$

where G denotes the induced inner product on the exterior algebra of vector fields and $\eta_1 \wedge \eta_2$ is the cross product of the normal distribution $\{\eta_1, \eta_2\}$ on $\partial\Sigma$.

Problem 9. *Optimize $J_\Sigma[\Gamma(\cdot)]$ subject to Equation (5).*

The corresponding Hamiltonian is:

$$
H^l(x, g, \Gamma, p) = X^l(x, g, \Gamma) - g^{is}\Gamma^j_{sk}p^{lk}_{ij}
$$

and the multivariate maximum principle is described in the following statement.

Corollary 8. *Suppose the linear connection $\Gamma^*(\cdot)$ is an optimal solution for $\left(\max\limits_{\Gamma(\cdot)} J_\Sigma[\Gamma(\cdot)], \text{Equation (5)}\right)$ and that $g^*(\cdot)$ is the corresponding optimal Riemannian structure. Then there exist a dual object $p^* = (p^{*lk}_{ij} = p^{*lk}_{ji} = -p^{*kl}_{ij})$ satisfying:*

- *State equations:*

$$
\dfrac{\partial g^{*ij}}{\partial x^k} = \dfrac{\partial H^l}{\partial p^{lk}_{ij}} + \dfrac{\partial H^l}{\partial p^{lk}_{ji}} \quad \text{(no sum on } l\text{)};
$$

- *Adjoint equations:*

$$
\left[\dfrac{\partial p^{*lk}_{ij}}{\partial x^k} + \left(\dfrac{\partial H^l}{\partial g^{ij}} + \dfrac{\partial H^l}{\partial g^{ji}}\right)\right]\dfrac{\partial}{\partial x^l} \in T_x\Sigma;
$$

- *Optimality conditions:*

$$
\left[\dfrac{\partial H^l}{\partial \Gamma^k_{ij}} + \dfrac{\partial H^l}{\partial \Gamma^k_{ji}}\right]\dfrac{\partial}{\partial x^l} \in T_x\Sigma;
$$

- *Boundary conditions:*

$$
G\left(\left[p^{*lk}_{ij} - \left(\dfrac{\partial \chi^{lk}}{\partial g^{ij}} + \dfrac{\partial \chi^{lk}}{\partial g^{ji}}\right)\right]\partial_{lk}, \eta_1 \wedge \eta_2\right) = 0.
$$

4.3. Riemannian Control with Curvilinear Integral Cost Functional

The natural expression for dual mapping necessary to phrase the optimality conditions requires curvature free Hamiltonians, therefore, if the cost functional is of a curvilinear type we can only analyze optimal control problems depending on connection. More precisely, we analyze the problem of optimizing $J_C[\Gamma(\cdot)]$ subject to Equation (5). The corresponding Hamiltonian 1-form is:

$$H_l(x, g, \Gamma, p) = X_l(x, g, \Gamma) - g^{is}\Gamma_{sl}^j p_{ij}$$

and the corresponding multivariate maximum principle is described by the following statement.

Corollary 9. *Suppose the linear connection $\Gamma^*(\cdot)$ is an optimal control solution for $\left(\max\limits_{\Gamma(\cdot)} J_C[\Gamma(\cdot)], \text{Equation (5)}\right)$ and that $g^*(\cdot)$ is the corresponding optimal Riemannian structure. Then there exist a dual tensor field $p^* = (p_{ij}^* = p_{ji}^*)$ to satisfy:*

- *The state equuations:*

$$\frac{\partial g^{*ij}}{\partial x^l} = \frac{\partial H_l}{\partial p_{ij}} + \frac{\partial H_l}{\partial p_{ji}};$$

- *the adjoint equations*

$$g^{ls}\left[\frac{\partial p_{ij}^*}{\partial x^l} - \left(\frac{\partial H_l}{\partial g^{ij}} + \frac{\partial H_l}{\partial g^{ji}}\right)\right]\frac{\partial}{\partial x^s} \in T_x^\perp C;$$

- *The optimality conditions:*

$$g^{ls}\left[\frac{\partial H_l}{\partial \Gamma_{ij}^k} + \frac{\partial H_l}{\partial \Gamma_{ji}^k}\right]\frac{\partial}{\partial x^s} \in T_x^\perp C;$$

- *The terminal conditions:*

$$p_{ij}^*(x) = \left[\frac{\partial \chi}{\partial g^{ij}} + \frac{\partial \chi}{\partial g^{ji}}\right](x), \ \forall x \in \{x_i, x_f\}.$$

5. Conclusions

The idea of finding optimal Riemannian structures for geometric meaningful integrals has classical roots. Nevertheless, the well-known Riemannian optimization approaches refer only to particular problems (like Hilbert's problem, or Plateau's problem) and the results are generally obtained via calculus of variations. This paper adapted multivariate optimal control techniques to general Riemannian optimization problems in order to derive a Hamiltonian approach. The cost functionals considered here were multiple, curvilinear, or hypersurface-type integrals. Descriptions for necessary optimality conditions were given. Furthermore, Hilbert's classical isoperimetric problem was solved in a Hamiltonian manner, together with another fresh example.

Author Contributions: The conceptualization, formal analysis, and validation of the content were done with equal participation of both authors.

Funding: This research was funded by Balkan Society of Geometers, Bucharest, Romania.

Acknowledgments: Many thanks to the Balkan Society of Geometers, who funded this research. The subject of this paper was scientifically supported by the University Politehnica of Bucharest, and by the Academy of Romanian Scientists.

Conflicts of Interest: The authors declare no conflict of interest.

References

1. Abbas, H.; Kim, Y.; Siegel, J.B.; Rizzo, D.M. Synthesis of Pontryagin's Maximum Principle Analysis for Speed Profile Optimization of All-Electric Vehicles. *J. Dyn. Sys. Meas. Control* **2019**, *141*, 071004. [CrossRef]

2. Ali, H.M.; Pereira, F.L.; Gama, S.M.A. A new approach to the Pontryagin maximum principle for nonlinear fractional optimal control problems. *Math. Methods Appl. Sci.* **2016**, *39*, 3640–3649. [CrossRef]

3. Avakov, E.R.; Magaril-Il'yaev, G.G. Generalized Maximum Principle in Optimal Control. *Doklady Math.* **2018**, *98*, 575–578. [CrossRef]

4. Ross, I.M. *A Primer on Pontryagin's Principle in Optimal Control*, 2nd ed.; Collegiate Publishers: San Frnacisco, CA, USA, 2015.

5. Udrişte, C. Minimal submanifolds and harmonic maps through multitime maximum principle. *Balkan J. Geom. Appl.* **2013**, *18*, 69–82.

6. Agrachev, A.; Sachkov, Y.L. Control theory from the geometric viewpoint. In *Encyclopedia of Mathematical Sciences*; Springer: Berlin, Germany, 2004; Volume 87.

7. Schattler, H.; Ledzewicz, U. Geometric Optimal Control: Theory, Methods and Examples. In *Interdisciplinary Applied Mathematics*; Springer: New York, NY, USA, 2012.

8. Zhu, J.; Trelat, E.; Cerf, M. Geometric optimal control and applications to aerospace. *Pac. J. Math. Ind.* **2017**, *9*, 8. [CrossRef]

9. Evans, L.C. *An Introduction to Mathematical Optimal Control Theory*; Lecture Notes; Department of Mathematics, University of California: Berkeley, CA, USA, 2010.

10. Baillieul, J.; Willems, J.C. *Mathematical Control Theory*; Springer Science and Business Media: New York, NY, USA, 1999.

11. Lee, E.B.; Markus, L. *Foundations of Optimal Control Theory*; Wiley: Hoboken, NJ, USA, 1967.

12. Macki, J.; Strauss, A. *Introduction to Optimal Control*; Springer: New York, NY, USA, 1982.

13. Pontriaguine, L.; Boltianski, V.; Gamkrelidze, R.; Michtchenko, E. *Théorie Mathématique des Processus Optimaux*; MIR: Moscow, Russia, 1974.

14. Cesari, L. Optimization with partial differential equations in Dieudonne-Rashevsky form and conjugate problems. *Arch. Rat. Mech. Anal.* **1969**, *33*, 339–357. [CrossRef]

15. Klotzler, R. On Pontrjagin's maximum principle for multiple integrals. *Beitrage zur Analysis* **1976**, *8*, 67–75.

16. Mititelu, Şt.; Pitea, A.; Postolache, M. On a class of multitime variational problems with isoperimetric constraints. *U.P.B. Sci. Bull. Ser. A Appl. Math. Phys.* **2010**, *72*, 31–40.

17. Pickenhain, S.; Wagner, M. Pontryaguin Principle for State-Constrained Control Problems Governed by First-Order PDE System. *J. Optim. Theory Appl.* **2000**, *107*, 297–330. [CrossRef]

18. Rund, H. Pontrjagin functions for multiple integral control problems. *J. Optim. Theory Appl.* **1976**, *18*, 511–520. [CrossRef]

19. Wagner, M. Pontryaguin Maximum Principle for Dieudonne-Rashevsky Type Problems Involving Lipcshitz functions. *Optimization* **1999**, *46*, 165–184. [CrossRef]

20. Kobayashi, S.; Nomizu, K. Foundations of Differential Geometry. In *Interscience Tracts in Pure and Applied Mathematics*; John Wiley and Sons Inc.: New York, NY, USA, 1963; Issue 15.

21. Lee, J.M. *Introduction to Riemannian Manifolds*, 2nd ed.; Graduate Texts in Mathematics; Springer: Cham, Switzerland, 2019.

22. Udrişte, C. Multitime stochastic control theory. In Proceedings of the Selected Topics on Circuits, Systems, Electronics, Control and Signal Processing, Cairo, Egypt, 29–31 December 2007; pp. 171–176.

23. Udrişte, C.; Ţevy, I. Multitime Dynamic Programming for Curvilinear Integral Actions. *J. Optim. Theory Appl.* **2010**, *146*, 189–207. [CrossRef]

24. Bejenaru, A.; Udrişte, C. Multitime optimal control and equilibrium deformations. In *Recent Researches in Hydrology, Geology and Continuum Mechanics*; WSEAS Press: Cambridge, UK, 2011; pp. 126–136.

25. Udrişte, C. Nonholonomic approach of multitime maximum principle. *Balkan J. Geom. Appl.* **2009**, *14*, 101–116.

26. Udrişte, C.; Ţevy, I. Multitime Dynamic Programming for Multiple Integral Actions. *J. Glob. Optim.* **2011**, *51*, 345–360. [CrossRef]

27. Udrişte, C.; Bejenaru, A. Multitime optimal control with area integral costs on boundary. *Balkan J. Geom. Appl.* **2011**, *16*, 138–154.

Symmetry **2019**, *11*, 893

28. Udrişte, C. Multitime controllability, observability and bang-bang principle. *J. Optim. Theory Appl.* **2008**, *139*, 141–157. [CrossRef]
29. Udrişte, C. Equivalence of multitime optimal control problems. *Balkan J. Geom. Appl.* **2010**, *15*, 155–162.

symmetry

MDPI

Article

Sasaki-Einstein 7-Manifolds, Orlik Polynomials and Homology

Ralph R. Gomez

Department of Mathematics and Statistics, Swarthmore College, Swarthmore, PA 19081, USA;
rgomez1@swarthmore.edu

Received: 4 June 2019; Accepted: 18 July 2019; Published: 23 July 2019

Abstract: In this article, we give ten examples of 2-connected seven dimensional Sasaki-Einstein manifolds for which the third homology group is completely determined. Using the Boyer-Galicki construction of links over particular Kähler-Einstein orbifolds, we apply a valid case of Orlik's conjecture to the links so that one is able to explicitly determine the entire third integral homology group. We give ten such new examples, all of which have the third Betti number satisfy $10 \leq b_3(L_f) \leq 20$.

Keywords: Sasaki-Einstein; Kähler 2; orbifolds; links

1. Introduction

A rich source of constructing Sasaki-Einstein (SE) metrics of positive Ricci curvature pioneered by Boyer and Galicki in Reference [1] is via links of isolated hypersurface singularities defined by weighted homogenous polynomials. These smooth manifolds have been used to show the existence of SE metrics on many types of manifolds such as exotic spheres [2], rational homology spheres ([3,4]) and connected sums of $S^2 \times S^3$ [1] (see Reference [5] for more comprehensive survey.) SE manifolds are also extremely important in relation to the AdS/CFT Correspondence which is a conjecture that, in certain environments, relates Sasaki-Einstein geometries to particular superconformal theories. (See for example, Reference [6] for recent progress in the relationship between SE geomtries and the AdS/CFT conjecture.) In general it is very difficult to determine the diffeomorphism or even homeomorphism type of a given link so determining any such geometric or topological data about the link is always helpful. Along these lines, for a given link of dimension $2n - 1$, Milnor and Orlik [7] determined a formula for the $n - 1$ Betti number of the link and later on Orlik conjectured a formula [8] (or see section two) for the torsion in $n - 1$ integral homology group. This conjecture due to Orlik regarding the torsion in integral homology of links is known to hold in certain cases. Both of these formulas have been instrumental in extracting some topological data on certain SE manifolds arising as links. For example, based on work of Cheltsov [9], Boyer gave fourteen examples [10] of SE 7-manifolds arising from links of isolated hypersurface singularities for which the third integral homology group is completely determined. He used Brieskorn-Pham polynomials and Orlik polynomials (see Section 1), both of which are cases in which the aforementioned conjecture holds. Inspired by these examples, the main motivation for this article is to find other examples of SE 7-manifolds arising as links generated by Brieskorn-Pham polynomials or Orlik polynomials so that one can explicitly calculate the third integral homology group.

In general, there are obstructions to finding SE metrics (e.g., Bishop obstruction and Lichnerowicz obstruction [11]) so it is worth finding as many examples as possible of manifolds which due admit SE metrics. Indeed, the main result of the paper is a list of ten examples (see Section 2) of SE links defined by Orlik polynomials. Because of this, we are then able to calculate the torsion in the third integral homology group explicitly. In Section 2, we review the necessary background and in Section 3 we give the table of ten examples together with the third Betti number and explicit forms of H_3.

2. Background

Define the weighted $\mathbb{C}^*$ action on $\mathbb{C}^{n+1}$ by

$$(z_0, ..., z_n) \longmapsto (\lambda^{w_0} z_0, ..., \lambda^{w_n} z_n)$$

where w_i are the weights which are positive integers and $\lambda \in \mathbb{C}^*$. We use the standard notation $\mathbf{w} = (w_0, ..., w_n)$ to denote a weight vector. In addition, we assume

$$gcd(w_0, ..., w_n) = 1.$$

Definition 1. *A polynomial $f \in \mathbb{C}[z_0, ..., z_n]$ is weighted homogenous if it satisfies*

$$f(\lambda^{w_0} z_0, ..., \lambda^{w_n} z_n) = \lambda^d f(z_0, ..., z_n)$$

for any $\lambda \in \mathbb{C}^*$ and the positive integer d is the degree of f.

The link L_f of an isolated hypersurface singularity defined by a weighted homogenous polynomial f with isolated singularity only at the origin is given by

$$L_f = C_f \cap S^{2n+1}$$

where C_f is the weighted affine cone defined by $f = 0$ in $\mathbb{C}^{n+1}$. By Milnor [12], L_f is a smooth $n - 2$ connected manifold of dimension $2n - 1$.

Recall a Fano orbifold Z is an orbifold for which the orbifold anticanonical bundle is ample.

Theorem 1 ([1]). *The link L_f as defined above admits as Sasaki-Einstein structure if and only if the Fano orbifold $\mathcal{Z}_f$ admits a Kähler-Einstein orbifold metric of scalar curvature $4n(n + 1)$*

Note that one simply needs to rescale a Kähler-Einstein metric of positive scalar curvature to get the desired scalar curvature in the statement of the theorem. We can think of the weighted hypersurface $\mathcal{Z}_f$ as the quotient space of the link L_f by the locally free circle action where this circle action comes from the weighted Sasakian structure on the link L_f. In fact this whole process is summarized in the commutative diagram [1]

$$
\begin{array}{ccc}
L_f & \longrightarrow & S^{2n+1} \\
\downarrow{\scriptstyle \pi} & & \downarrow \\
\mathcal{Z}_f & \longrightarrow & \mathbb{P}(\mathbf{w})
\end{array}
$$

where $S_{\mathbf{w}}^{2n+1}$ denotes the unit sphere with a weighted Sasakian structure, $\mathbb{P}(\mathbf{w})$ is weighted projective space coming from the quotient of $S_{\mathbf{w}}^{2n+1}$ by a weighted circle action generated from the weighted Sasakian structure. The top horizontal arrow is a Sasakian embedding and the bottom arrow is Kähler embedding. Moreover the vertical arrows are orbifold Riemannian submersions.

Thus, a mechanism for constructing 2-connected Sasaki-Einstein 7-manifolds boils down to finding orbifold Fano Kähler-Einstein hypersurfaces in weighted projective 4-space $\mathbb{P}(\mathbf{w})$. Johnson and Kollár in Reference [13] construct 4442 Fano orbifolds and of this list, 1936 of these are known to admit orbifold Kähler-Einstein metrics. Therefore, by the above construction we state a theorem of Boyer, Galicki and Nakamaye:

Theorem 2 ([3]). *There exists 1936 2-connected Sasaki-Einstein 7-manifolds realized as links of isolated hypersurface singularities defined by weighted homogenous polynomials.*

In Reference [3], the authors were able to determine many from the list of 1936 which yield rational homology 7-spheres and they also determined the order of $H_3(L_f, \mathbb{Z})$. In this paper, we identify ten links of isolated hypersurface singularities which can be given by so called Orlik polynomials, thus allowing us to calculate the third integral homology group explicitly. First, we need to define some quantities [7]:

$$u_i = \frac{d}{gcd(d, w_i)}, \qquad v_i = \frac{w_i}{gcd(d, w_i)}$$

Let L_f denote a link of an isolated hypersurface singularity defined by a weighted homogenous polynomial. The formula for the Betti number $b_{n-1}(L_f)$ is given by:

$$b_{n-1}(L_f) = \sum (-1)^{n+1-s} \frac{u_{i_1}, \ldots u_{i_s}}{v_{i_1} \cdots v_{i_s} lcm(u_{i_1}, \ldots, u_{i_s})}.$$

Here the sum is over all possible 2^{n+1} subsets $\{i_1, \ldots, i_s\}$ of $\{0, \ldots n\}$.

For the torsion data, Orlik conjectured [8] that for a given link L_f of dimension $2n - 1$ one has

$$H_{n-1}(L_f, \mathbb{Z})_{tor} = \mathbb{Z}_{d_1} \oplus \mathbb{Z}_{d_2} \oplus \cdots \oplus \mathbb{Z}_{d_r} \tag{1}$$

We should now review how the d_i data are given, using the presentation given in Reference [5]. Given an index set $\{i_1, i_2, \ldots, i_s\}$, define I to be the set of all of the 2^s subsets and let us designate J to be all of the proper subsets. For each possible subset, we must define (inductively) a pair of numbers $c_{i_1, \ldots, i_s}$ and $k_{i_1, \ldots, i_s}$. For each ordered subset $\{i_1, \ldots, i_s\} \subset \{0, 1, 2, \ldots, n\}$ with $i_1 < i_2 < \cdots < i_s$ one defines the set of 2^s positive integers, beginning with $c_{\varnothing} = gcd(u_0, \ldots, u_n)$:

$$c_{i_1, \ldots, i_s} = \frac{gcd(u_0, \ldots, \widehat{u}_{i_1}, \ldots, \widehat{u}_{i_s}, \ldots, u_n)}{\prod_J c_{j_1, \ldots, j_t}}.$$

Now, to get the k's:

$$k_{i_1, \ldots, i_s} = \epsilon_{n-s+1} \kappa_{i_1, \ldots, i_s} = \epsilon_{n-s+1} \sum_I (-1)^{s-t} \frac{u_{j_1} \cdots u_{j_t}}{v_{j_1} \cdots v_{j_t} lcm(u_{j_1}, \ldots, u_{j_t})}$$

where

$$\epsilon_{n-s+1} = \begin{cases} 0, & \text{if } n - s + 1 \text{ is even} \\ 1, & \text{if } n - s + 1 \text{ is odd}. \end{cases}$$

Then for each $1 \le j \le r = \lfloor max\{k_{i_1, \ldots, i_s}\} \rfloor$ we put

$$d_j = \prod_{k_{i_1, \ldots, i_s} \ge j} c_{i_1, \ldots, i_s}.$$

Though the full conjecture is still open 45 years later, it is known to hold in certain cases. If the link is given by either of the polynomials below

$$z_0^{a_0} + z_1^{a_1} + \cdots + z_n^{a_n}, \qquad z_0^{a_0} + z_0 z_1^{a_1} + \cdots + z_{n-1} z_n^{a_n} \tag{2}$$

then the conjecture holds [8,14]. The first type of polynomial is called Brieskorn-Pham and the second one is called Orlik. We will discuss these a bit more in the next section.

The formulas for the Betti numbers and torsion would indeed be quite tedious to compute by hand, especially when the degree and the weights are large. Fortunately, Evan Thomas developed a program written in C which computes the Betti numbers and the numbers d_i, which generates the torsion in $H_{n-1}(L_f, \mathbb{Z})$. Hence if the link is generated by a Brieskorn-Pham polynomial or an Orlik polynomial, then one explicitly knows the torsion in H_{n-1}. This program was also used extensively in

References [5,10,15]. I would like to thank Evan Thomas for giving me permission to use the program and to make it available. See the Appendix A.

3. Examples

The paper of Johnson and Kollár [13] lists (see appendix for link to list) Kähler-Einstein and Tiger of Fano orbifolds in weighted projective space $\mathbb{P}(\mathbf{w})$ gives the weight vector $\mathbf{w} = (w_0, w_1, w_2, w_3, w_4)$ with $w_0 \leq w_1 \leq w_2 \leq w_3 \leq w_4$ (which can always be done after an affine change of coordinates) and it indicates if the weighted hypersurface admits an orbifold Kähler-Einstein structure. The degree d is given by $d = (w_0 + \cdots + w_4) - 1$. It is easy to identify whether or not Kähler-Einstein orbifolds on the list come from Brieskorn-Pham polynomials since for a given weight vector $\mathbf{w}$ on the list, the exponents of the Brieskorn-Pham polynomial would have to be $a_i = d/w_i$ for $i = 0, ..., 4$ and therefore one can do a computer search to see if one gets integer results for the exponents. But are there any coming from Orlik polynomials? To get some Orlik examples, one must search among the weighted hypersurfaces in the list of 1936 Kähler-Einstein orbifolds and see if the given weights can be represented by Orlik polynomials. This is more difficult than in the Brieskorn-Pham case since the constraints, given in 3.1, are more complicated. The search was done within the range $9 \leq w_0 \leq 11$ where there are 436 Fano orbifolds. Of this lot, 149 Fano orbifolds are known to admit an orbifold Fano Kähler-Einstein structure. Therefore, for a given weight vector $\mathbf{w} = (w_0, w_1, w_2, w_3, w_4)$ one needs to see if there exists exponents a_i, in the Orlik polynomials satisfying

$$d = a_0 w_0 = w_0 + w_1 a_1 = w_1 + w_2 a_2 = w_2 + w_3 a_3 = w_3 + w_4 a_4. \tag{3}$$

The ten examples were found by hand, checking many different weights against the given conditions. Once they were found, the computer program developed by Evan Thomas was implemented to determine the Betti number and the torsion data. We now give the table of ten examples. We list the weights, the quasihomogenous polynomial generating the link, the degree and finally the third homology group. It is not claimed that this list is exhaustive. There may very well be more examples using these methods.

$(75,10,163,331,247)$	$z_0^{11}+z_0 z_1^{75}+z_1 z_2^5+z_2 z_3^2+z_3 z_4^2$	825	$\mathbb{Z}^{10} \oplus \mathbb{Z}_{55} \oplus (\mathbb{Z}_5)^4$
$(62,124,155,9,85)$	$z_0^7+z_0 z_1^3+z_1 z_2^2+z_2 z_3^{31}+z_3 z_4^5$	434	$\mathbb{Z}^{12} \oplus \mathbb{Z}_{14} \oplus (\mathbb{Z}_2)^2$
$(9,174,467,277,649)$	$z_0^{175}+z_0 z_1^2+z_1 z_2^3+z_2 z_3^4+z_3 z_4^2$	1575	$\mathbb{Z}^{12} \oplus \mathbb{Z}_{525} \oplus (\mathbb{Z}_3)^2$
$(87,348,145,11,193)$	$z_0^9+z_0 z_1^2+z_1 z_2^3+z_2 z_3^{58}+z_3 z_4^4$	783	$\mathbb{Z}^{12} \oplus \mathbb{Z}_{27} \oplus \mathbb{Z}_3$
$(100,350,9,113,229)$	$z_0^8+z_0 z_1^2+z_1 z_2^{50}+z_2 z_3^2+z_3 z_4^3$	800	$\mathbb{Z}^{14} \oplus \mathbb{Z}_{400}$
$(9,291,488,181,787)$	$z_0^{195}+z_0 z_1^6+z_1 z_2^3+z_2 z_3^7+z_3 z_4^2$	1755	$\mathbb{Z}^{14} \oplus \mathbb{Z}_{585} \oplus \mathbb{Z}_3$
$(10,164,333,71,253)$	$z_0^{83}+z_0 z_1^5+z_1 z_2^2+z_2 z_3^4+z_3 z_4^3$	830	$\mathbb{Z}^{14} \oplus \mathbb{Z}_{166}$
$(10,540,275,163,103)$	$z_0^{109}+z_0 z_1^2+z_1 z_2^2+z_2 z_3^5+z_3 z_4^9$	1090	$\mathbb{Z}^{16} \oplus \mathbb{Z}_{218} \oplus \mathbb{Z}_2$
$(32,144,11,103,31)$	$z_0^{10}+z_0 z_1^2+z_1 z_2^{16}+z_2 z_3^3+z_3 z_4^4$	320	$\mathbb{Z}^{18} \oplus \mathbb{Z}_{160}$
$(45,36,27,11,107)$	$z_0^5+z_0 z_1^5+z_1 z_2^2+z_2 z_3^{18}+z_3 z_4^2$	225	$\mathbb{Z}^{20} \oplus \mathbb{Z}_5$

4. Conclusions

Because Sasaki-Einstein manifolds of positive Ricci curvature play such an important role in the AdS/CFT conjecture in string theory, it is of utmost importance to have as many examples as possible of Sasaki-Einstein manifolds especially in dimensions five and seven. In this paper, ten new examples of seven dimensional 2-connected Sasaki-Einstein manifolds were constructed by constructing links using Orlik polynomials over particular Kähler-Einstein Fano orbifolds. The third homology group was explicitly calculated using a conjectural formula which is known to be true for Orlik polynomials. It is likely there are more examples using this approach but is difficult to detect them without a more systematic approach.

Funding: Part of this article was prepared with the support of the James Michener Fellowship of Swarthmore College.

Acknowledgments: I would like to thank Charles Boyer for useful conversations and I would like to express my gratitude to Evan Thomas for allowing me to use the program he developed and for giving me permission to share the code.

Conflicts of Interest: The author declares no conflict of interest.

Appendix A

(a) The Johnson-Kollár list of hypersurfaces in weighted projective 4-space $\mathbb{P}(\mathbf{w})$ admitting Kähler-Einstein orbifold metrics is available at https://web.math.princeton.edu/~jmjohnso/delpezzo/ KEandTiger.txt. It lists the weights followed by data on wether or not it is known if the hypersurface admits a Kähler-Einstein orbifold metric.

(b) The code developed by Evan Thomas to compute the homology of links is available at https: //blogs.swarthmore.edu/gomez/wp-content/uploads/2016/07/evans.c.

References

1. Boyer, C.P.; Galicki, K. New Einstein Metrics in Dimension Five. *J. Differ. Geom.* **2001**, *57*, 443–463. [CrossRef]
2. Boyer, C.P.; Galikci, K.; Kollár, J. Einstein Metrics on Spheres. *Ann. Math.* **2005**, *162*, 557–580. [CrossRef]
3. Boyer, C.P.; Galicki, K.; Nakamaye, M. Einstein Metrics on Rational Homology 7-Spheres. *Ann. Inst. Fourier* **2002**, *52*, 1569–1584. [CrossRef]
4. Boyer, C.P.; Galicki, K.; Nakamaye, M. Einstein Metrics on Rational Homology Spheres. *J. Differ. Geom.* **2006**, *74*, 353–362. [CrossRef]
5. Boyer, C.P.; Galicki, K. *Sasakian Geometry*; Oxford Mathematical Monographs; Oxford University Press: Oxford, UK, 2007.
6. Xie, D.; Shing-Tung, Y. Singularity, Sasaki-Einstein Manifold, Log del Pezzo Surfaces and N=1 AdS/CFT correspondence: Part 1. Available online: https://arxiv.org/abs/1903.00150 (accessed on 4 June 2019).
7. Milnor, J.; Orlik, P. Isolated Singlarities Defined by Weighted Homogeneous Polynomials. *Topology* **1970**, *9*, 385–393. [CrossRef]
8. Orlik, P. On the Homology of Weighted Homogenous Manifolds. In *Proceedings of the Second Conference on Compact Transformation Groups*; (Univ. Mass, Amherst, Mass 1971) Part I (Berlin); Spring: Berlin/Heidelberg, Germany, 1972; pp. 260–269.
9. Cheltsov, I. Fano Varieties With Many Self-Maps. *Adv. Math.* **2008**, *217*, 97–124. [CrossRef]
10. Boyer, C.P. Sasakian Geometry: The Recent Work of Krzysztof Galicki. *Note di Matematica* **2008**, *28*, 63–105.
11. Gauntlett, J.; Martelli, D.; Sparks, J. Obstructions to the Existence of Sasaki-Einstein Metrics. *Commun. Math. Phys.* **2006**, *273*, 803–827. [CrossRef]
12. Milnor, J. *Singular Points of Complex Hypersurfaces*; Annals of Mathematical Studies, Volume 61; Princeton University Press: Princeton, NJ, USA, 1968.
13. Johnson, J.M.; Kollár, J. Fano Hypersurfaces in Weighted Projective 4-Space. *Exp. Math.* **2001**, *10*, 151–158. [CrossRef]
14. Orlik, P.; Randall, R. The Monodromy of Weighted Homogenous Singularities. *Invent. Math.* **1977**, *39*, 199–211. [CrossRef]
15. Boyer, C.P.; Galicki, K. Sasakian Geometry, Hypersurface Singularities, and Einstein Metrics. *Supplemento ai Rendiconti del Circolo Matematico di Palermo Serie II* **2005**, *75* (Suppl.), 57–87.

symmetry

MDPI

Article

On Formality of Some Homogeneous Spaces

Aleksy Tralle

Faculty of Mathematics and Computer Science, University of Warmia and Mazury, Słoneczna 54,
10-710 Olsztyn, Poland; tralle@matman.uwm.edu.pl

Received: 17 July 2019; Accepted: 2 August 2019; Published: 5 August 2019

Abstract: Let G/H be a homogeneous space of a compact simple classical Lie group G. Assume that the maximal torus T_H of H is conjugate to a torus T_β whose Lie algebra $\mathfrak{t}_\beta$ is the kernel of the maximal root β of the root system of the complexified Lie algebra $\mathfrak{g}^c$. We prove that such homogeneous space is formal. As an application, we give a short direct proof of the formality property of compact homogeneous 3-Sasakian spaces of classical type. This is a complement to the work of Fernández, Muñoz, and Sanchez which contains a full analysis of the formality property of $SO(3)$-bundles over the Wolf spaces and the proof of the formality property of homogeneous 3-Sasakian manifolds as a corollary.

Keywords: formality; 3-Sasakian manifold; homogeneous space

1. Introduction

Formality is an important homotopic property of topological spaces. It is often related to the existence of particular geometric structures on manifolds. For example, Kaehler manifolds are formal [1], and the same holds for compact Riemannian symmetric spaces [2,3]. In general, Sasakian manifolds do not possess this property. However, their higher order Massey products vanish [4], and this can be regarded as a "formality-like" property as well. An interesting issue is the formality of homogeneous spaces of compact Lie groups. For example, Amann [5] found several characterizations of non-formality of homogeneous spaces. Some homogeneous spaces determined by characters of maximal tori are not formal [6,7]. On the other hand, compact homogeneous spaces of positive Euler characteristics are known to be formal [3,7] and the same holds for G/H generated by a finite order automorphism of G [8]. It should be noted that there is a general method of studying the formality property of homogeneous spaces in terms of the Lie group-theoretic data [3,7]. However, such methods may work for a *given* pair (G, H) together with the *known* embedding of H into G. Hence, it is still interesting to find geometrically important classes of homogeneous spaces satisfying formality or non-formality property. In this article, we prove the following result.

Theorem 1. *Let G/H be a homogeneous space of a compact simple classical Lie group G. Assume that the maximal torus T_H of H is conjugate (in G) to the torus T_β whose Lie algebra is the kernel $\mathrm{Ker}\ \beta$ of the maximal root β of the root system $\Delta(\mathfrak{g}^c)$. Then G/H is formal.*

This class of homogeneous spaces has geometric significance. To show this we present the following geometric application. In [9] the formality property of $SO(3)$-bundles over the Wolf spaces was analyzed. Consequently, one obtains the formality property of any compact homogeneous 3-Sasakian manifold. In this note we show that if one restricts himself to this class of Riemannian manifolds, then the proof can be obtained entirely in terms of the data of the 3-Sasakian homogeneous space G/H (at least for classical Lie groups G). Thus, we give a direct proof the following result [9].

Theorem 2. *Let G be a classical compact simple Lie group. Then, any 3-Sasakian homogeneous space G/H is formal.*

Although [9] contains much stronger and more general result, the direct proof still may be of independent interest. This is motivated by the fact that homogeneous 3-Sasakian manifolds G/H admit a description in terms of the root systems of the complexified Lie algebra $\mathfrak{g}^c$, and in some cases, the formality property can be expressed via the same data [7] (see also [5,6]). It seems to make a remark that Theorem 1 probably holds for all simple Lie groups. However, the method of proof uses the generators of the ring of invariants of the Weyl group, which becomes computationally difficult (compare, for example the expressions of such polynomials for the exceptional Lie groups [10]).

2. Preliminaries

2.1. Presentation and Notation

We approach the problem of formality from the point of view of the classical cohomology theory of homogeneous spaces of compact Lie groups [7,11]. We use the basic notions and facts from the theory of Lie groups and Lie algebras without explanations. Instead, we refer to [12]. We denote Lie groups by capital letters $G, H, \ldots$, and their Lie algebras by the corresponding Gothic letters $\mathfrak{g}, \mathfrak{h}, \ldots$. Let G be a compact semisimple Lie group. The real cohomology algebra $H^*(G)$ is isomorphic to the exterior algebra over the space of primitive elements $P_G = \langle y_1, \ldots, y_n \rangle$:

$$H^*(G) \cong \Lambda\, P_G = \Lambda(y_1, \ldots, y_n),\, y_i \in P_G, i = 1, \ldots, n = \operatorname{rank}\, G.$$

The degrees of y_i are equal to $2p_i - 1$, where p_i are the exponents of $\mathfrak{g}$. We denote by S_G the ring of G-invariant polynomials on the Lie algebra $\mathfrak{g}$. Let T be a maximal torus of G. Consider the Weyl group $W_G = N_G(T)/T$. It acts on $\mathfrak{t}$ and on the polynomial algebra $\mathbb{R}[\mathfrak{t}]$ of all polynomials over $\mathfrak{t}$. The subring S_{W_G} of W_G-invariants in $\mathbb{R}[\mathfrak{t}]$ is generated by $n = \operatorname{rank}\, G$ polynomials $F_1, \ldots, F_n$ of degrees $2p_i$. The following isomorphism is well known [7,11]:

$$S_G \cong S_{W_G} \cong \mathbb{R}[\mathfrak{t}]^{W_G} \cong \mathbb{R}[F_1, \ldots, F_n].$$

We will use a map $\tau_G : \Lambda\, P_G \to S_G$ called the *transgression map* [7,11]. The transgression τ_G maps y_i, $i = 1, \ldots, n$ onto some free generators of S_{W_G}. We follow [9] in the presentation of Sasakian and 3-Sasakian manifolds. One can also consult [13].

2.2. Formality

Here we recall some definitions and facts from the theory of minimal models and formality [14].

We consider *differential graded commutative algebras*, or DGAs, over the field $\mathbb{R}$ of real numbers. The degree of an element a of a DGA is denoted by $|a|$.

Definition 1. *A DGA $(\mathcal{A}, d)$ is minimal if:*

1. *$\mathcal{A}$ is the free algebra $\wedge V$ over a graded vector space $V = \bigoplus_i V^i$, and*
2. *there is a family of generators $\{a_\tau\}_{\tau \in I}$ indexed by some well-ordered set I, such that $|a_\mu| \leq |a_\tau|$ if $\mu < \tau$ and each da_τ is expressed in terms of preceding a_μ, $\mu < \tau$. Thus, da_τ does not have a linear part.*

An important example of DGA is the de Rham algebra $(\Omega^*(M), d)$ of a differentiable manifold M, where d is the exterior differential. This DGA will be used in this article.

Given a differential graded commutative algebra $(\mathcal{A}, d)$, we denote its cohomology by $H^*(\mathcal{A})$. The cohomology of a differential graded algebra $H^*(\mathcal{A})$ is also a DGA with the multiplication inherited from that on $\mathcal{A}$ and with zero differential. The DGA $(\mathcal{A}, d)$ is *connected* if $H^0(\mathcal{A}) = \mathbb{R}$, and $\mathcal{A}$ is *1-connected* if, in addition, $H^1(\mathcal{A}) = 0$. Morphisms between DGAs are required to preserve the degree and to commute with the differential.

Definition 2. *A free graded differential algebra* $(\bigwedge V, d)$ *is called a minimal model of the differential graded commutative algebra* $(\mathcal{A}, d)$ *if* $(\bigwedge V, d)$ *is minimal and there exists a morphism of differential graded algebras*

$$\rho : (\bigwedge V, d) \longrightarrow (\mathcal{A}, d)$$

inducing an isomorphism $\rho^* : H^*(\bigwedge V) \xrightarrow{\sim} H^*(\mathcal{A})$ *of cohomologies.*

Definition 3. *Two DGAs* $(\mathcal{A}, d_A)$ *and* $(\mathcal{B}, d_B)$ *are quasi-isomorphic, if there is a sequence of DGA algebras* $(\mathcal{A}_i, d_i)$ *and a sequence of morphisms between* $(\mathcal{A}_i, d_i)$ *and* $(\mathcal{A}_{i+1}, d_{i+1})$ *with* $(\mathcal{A}_1, d_1) = (\mathcal{A}, d_A)$ *and* $(\mathcal{A}_n, d_n) = (\mathcal{B}, d_B)$ *such that these morphisms induce isomorphisms of the corresponding cohomology algebras (the morphisms may be directed arbitrarily).*

It is known [14] that any connected differential graded algebra $(\mathcal{A}, d)$ has a minimal model which is unique up to isomorphism.

Definition 4. *A minimal model of a connected differentiable manifold M is a minimal model* $(\bigwedge V, d)$ *for the de Rham complex* $(\Omega^*(M), d)$ *of differential forms on M.*

If M is a simply connected manifold, then the dual $(\pi_i(M) \otimes \mathbb{R})^*$ of the vector space $\pi_i(M) \otimes \mathbb{R}$ is isomorphic to V^i for any i. This duality shows the relation between minimal models and homotopy groups. The same result is valid when $i > 1$, the fundamental group $\pi_1(M)$ is nilpotent and its action on $\pi_j(M)$ is nilpotent for all $j > 1$.

Definition 5. *A minimal algebra* $(\bigwedge V, d)$ *is called formal if there exists a morphism of differential algebras* $\psi : (\bigwedge V, d) \longrightarrow (H^*(\bigwedge V), 0)$ *inducing the identity map on cohomology.*

A smooth manifold M is called *formal* if its minimal model is formal. Examples of formal manifolds are ubiquitous: spheres, projective spaces, compact Lie groups, some homogeneous spaces, flag manifolds, and all compact Kaehler manifolds [1,3,5,8,14].

It is important to note that quasi-isomorphic minimal algebras have isomorphic minimal models. Therefore, to study formality of manifolds, one can use other "algebraic models". This means that one may take any DGAs $(\mathcal{A}, d_A)$ which are quasi-isomorphic to the de Rham algebra. This will be used in our analysis of formality of homogeneous spaces.

2.3. Quaternionic-Kaehler and 3-Sasakian Manifolds

A Riemannian $4n$-dimensional manifold (X, h) is called *quaternionic-Kaehler*, if the holonomy group $\mathrm{Hol}(X, h)$ is contained in $Sp(n)Sp(1)$.

An odd dimensional Riemannian manifold (M, g) is Sasakian if its cone $(M \times \mathbb{R}^+, g^c = t^2 g + dt^2)$ is Kaehler. This means that there is a compatible integrable almost complex structure J so that $(M \times \mathbb{R}^+, g^c, J)$ is a Kaehler manifold. In this case, the vector field $\xi = J \frac{\partial}{\partial t}$ is a Killing vector field of unit length. The 1-form η defined by $\eta(X) = g(\xi, X)$ for any vector field X on M is a contact form, whose Reeb vector field is ξ. Let ∇ denote the Levi-Civita connection of g. The $(1, 1)$-tensor $\phi(X) = \nabla_X \xi$ satisfies the identities

$$\phi^2 = -\operatorname{id} + \eta \otimes \xi, \; g(\phi(X), \phi(Y)) = g(X, Y) - \eta(X)\eta(Y),$$

$$d\eta(X, Y) = 2g(\phi(X), Y),$$

for any vector fields X, Y.

A Riemannian manifold (M, g) of dimension $4n + 3$ is called 3-*Sasakian*, if the cone $(M \times \mathbb{R}^+, g^c)$ admits three compatible integrable almost complex structures J_1, J_2, J_3 such that

$$J_1 J_2 = -J_2 J_1 = J_3,$$

and such that $(M \times \mathbb{R}^+, g^c, J_1, J_2, J_3)$ is a hyperkaehler manifold. Thus, (M, g) admits three Sasakian structures with Reeb vector fields ξ_1, ξ_2, ξ_3 of the contact forms η_1, η_2, η_3, and three tensors ϕ_1, ϕ_2, ϕ_3. The following relations are satisfied:

$$\eta_i(\xi_j) = g(\xi_i, \xi_j) = \delta_{ij}, \phi_i(\xi_j) = -\phi_j(\xi_i) = \xi_k,$$

$$\eta_i \circ \phi_j = -\eta_j \circ \phi_i = \eta_k$$

$$\phi_i \circ \phi_j - \eta_j \otimes \xi_i = -\phi_j \circ \phi_i + \eta_i \otimes \xi_j = \phi_k,$$

$$[\xi_i, \xi_j] = 2\xi_k,$$

for any cyclic permutation of (i, j, k) of $(1, 2, 3)$.

Let (M, g) be a Riemannian manifold carrying a 3-Sasakian structure. Denote by $\mathrm{Aut}(M, g)$ the subgroup of the isometry group $\mathrm{Iso}(M, g)$ consisting of all isometries preserving the 3-Sasakian structure

$$(g, \xi_s, \eta_s, \phi_s, s = 1, 2, 3).$$

By definition, a 3-Sasakian manifold (M, g) is called *homogeneous*, if $\mathrm{Aut}(M, g)$ acts transitively on M.

By definition, a *Wolf space* is a homogeneous quaternionic-Kaehler manifold of positive scalar curvature. The classification of the Wolf spaces is known [15,16] and can be reproduced as follows:

$$\mathbb{HP}^n = Sp(n+1)/(Sp(n) \times Sp(1)), \, \mathbb{Gr}_2(\mathbb{C}^{n+2}), \, \mathbb{Gr}_4(\mathbb{R}^{n+4}),$$

$$GI = G_2/SO(4), \, FI = F_4/Sp(3) \cdot Sp(1), \, EII = E_6/SU(6) \cdot Sp(1),$$

$$EVI = E_7/Spin(12) \cdot Spin(1), \, EIX = E_8/E_7 \cdot Sp(1).$$

Here $\mathbb{Gr}_4(\mathbb{R}^{n+4})$ denotes the Grassmannian of oriented real 4-planes. It follows that the classification of homogeneous 3-Sasakian manifolds is given by the following result (see [9], Section 2).

Theorem 3. *Let (M, g) be a 3-Sasakian homogeneous space. Then M is the total space of the fiber bundle*

$$F \to M \to W$$

over a Wolf space W. The fiber F is $Sp(1)$ for $M = S^{4n+3}$ and it equals $SO(3)$ in all other cases. Moreover, M is the one of the following homogeneous spaces:

$$Sp(n+1)/Sp(n) \cong S^{4n+3}, \, Sp(n+1)/(Sp(n) \times \mathbb{Z}_2),$$

$$SU(n+2)/S(U(n) \times U(1)), SO(m+4)/SO(m) \times Sp(1),$$

$$G_2/Sp(1), F_4/Sp(3), E_6/SU(6), E_7/Spin(12), E_8/E_7,$$

where $k \geq 0, n \geq 1, m \geq 3$. For the first two cases $Sp(0)$ means the trivial group.

3. Proof of Theorem 1

3.1. A Theorem on Formality of Homogeneous Spaces

Theorem 4 ([5]). *Let G/H be a homogeneous space of a compact semisimple Lie group G and let T_H be a maximal torus in H. Then G/H is formal if and only G/T_H is formal.*

3.2. Cartan Algebras

The material of this subsection is presented following [7]. It is well known that a homogeneous space G/H of a compact semisimple Lie group G has an algebraic model (which is called the Cartan algebra) of the form

$$(C(\mathfrak{g}, \mathfrak{h}), d) = (S_H \otimes \Lambda P_G, d)$$

where

$$d(q \otimes 1) = 0, \ \forall q \in S_H$$

$$d(1 \otimes p) = j^*(\tau_G(p)), \ \forall p \in \Lambda P_G.$$

Here $\tau_G : \Lambda P_G \to S_G$ is the transgression, $j^* : S_G \to S_H$ is a restriction map, and S_G, S_H are the algebras of invariant polynomials on $\mathfrak{g}$ and $\mathfrak{h}$, respectively. In particular, if $H = T$ for some torus in G, then j^* is a restriction of any invariant polynomial in S_G onto the Lie algebra $\mathfrak{t}$. Please note that T need not be maximal.

More generally, consider the DGA algebra of the form

$$(C, d) = (\mathbb{R}[x_1, \ldots, x_m] \otimes \Lambda(y_1, \ldots, y_n), d)$$

with the differential d vanishing on x_i, $i = 1, \ldots, m$ and

$$d(y_j) = F_j(x_1, \ldots, x_m).$$

We assume that y_j have some odd degrees $2l_j - 1$. Let $H^*(C)$ be the cohomology algebra of (C, d). We will also use the notation

$$H^*(C) = H(F_1, \ldots, F_n)$$

to stress the role of the ideal $I = (F_1, \ldots, F_n)$ (in the polynomial ring $\mathbb{R}[x_1, \ldots, x_m]$).

Recall the following definition. Let A be any commutative ring. A sequence $a_1, \ldots, a_k$ of elements in A is called *regular*, if a_i is not a zero divisor in $A/(a_1, \ldots, a_{i-1})$.

The following characterization of formality of a general Cartan algebra (C, d) is well known [7].

Theorem 5. *A general Cartan algebra (C, d) is formal if and only if the ideal $(F_1, \ldots, F_n)$ has the following property: the minimal system of generators is regular. The number of such generators cannot exceed m.*

Finally, recall the following isomorphism

$$S_G \cong S_{W_G} \cong \mathbb{R}[\mathfrak{t}]^{W_G},$$

where S_{W_G} denotes the ring of polynomials on $\mathfrak{t}$ which are invariant with respect to the action of the Weyl group W_G of G. Also, there is a commutative diagram

$$\begin{array}{ccc} S_G & \longrightarrow & S_{W_G} \\ j^* \downarrow & & j^* \downarrow \\ S_H & \longrightarrow & S_{W_H} \end{array}$$

which shows that the Cartan algebra $(C(\mathfrak{g}, \mathfrak{h})$ is isomorphic to the general Cartan algebra of the form

$$(C, d) = \mathbb{R}[\mathfrak{t}_H]^{W_H} \otimes \Lambda(y_1, \ldots, y_n)$$

$$d(y_k) = j^*(F_k), k = 1, \ldots, n, F_k \in \mathbb{R}[\mathfrak{t}]^{W_G}.$$

Here F_k are free generators of the ring of invariants $\mathbb{R}[\mathfrak{t}]^{W_G}$ determined by the transgression.

Please note that in the sequel we will use the particular choices of free invariant generators of polynomial algebras $\mathbb{R}[\mathfrak{t}]^{W_G}$ for each simple compact Lie group. These can be found in many sources, we use [7], Example 1 on page 186.

3.3. Formality of G/T_β

Proposition 1. *Let G/T_β be a homogeneous space of a compact classical Lie group G and a torus T_β whose Lie algebra is the kernel of the maximal root. Then G/T_β is formal.*

Proof. The proof is based on the checking of the conditions of Theorem 5 for G/T_β in each case A_n, B_n, C_n, D_n separately (although the calculations are very similar). Also, due to the final remark in the previous section, we can consider the algebraic model of G/T_β in the form

$$(\mathbb{R}[\mathfrak{t}_\beta] \otimes \Lambda(y_1, \ldots, y_n), d)$$

with

$$d(y_i) = F_i|_{\mathfrak{t}_\beta}, i = 1, \ldots, n.$$

In the proof we use the description of the maximal roots of the root systems of classical type [15].

Case 1 (C_n). In this case, in the coordinates $x_1, \ldots, x_n$ in $\mathfrak{t}$, the maximal root β has the form $\beta = 2x_1$. Thus, $\mathfrak{t}_\beta$ is determined by the equation $x_1 = 0$, and the restrictions of F_i on $\mathfrak{t}_\beta$ have the form $F_i|_{\mathfrak{t}_\beta} = F_i(0, x_2, \ldots, x_n)$. Please note that the ring of invariants $\mathbb{R}[\mathfrak{t}]^{W_G}$ may have different sets of generators, and in general we cannot take them arbitrarily, because they are determined by the transgression. However, by Theorem 5, *the formality property is determined not by the particular polynomials, but by the whole ideal* $(F_1, \ldots, F_n)$. It follows that one can work with any set of generators. In case of C_n we can take

$$F_i(x_1, \ldots, x_n) = x_1^{2i} + \cdots + x_n^{2i}, i = 1, \ldots, n.$$

The restrictions onto $\mathfrak{t}_\beta$ have the form

$$F_i(0, x_2, \ldots, x_n),$$

this sequence is obviously regular for $i = 1, \ldots, n-1$. Since the number of variables is also $n-1$, the result follows.

Case 2 (B_n). Here $\beta = x_1 + x_2$. We make the same argument to the previous case. Again, one may choose the invariant generators in the form $F_i = \sum_{k=1}^{n} x_k^{2i}, i = 1, \ldots, n$. This time the restrictions will take the form

$$F_i|_{\mathfrak{t}_\beta} = F_i(-x_2, x_2, x_3 \ldots, x_n) = 2x_2^{2i} + x_3^{2i} + \cdots + x_n^{2i}.$$

Again, this sequence is obviously regular for $i = 1, \ldots, n-1$ and the result follows from Theorem 5.

Case 3 (D_n). In this case, again, $\beta = x_1 + x_2$. However, the invariant generators are different. One of the possible choices is

$$F_i(x_1, \ldots, x_n) = \sum_{k=1}^{n} x_k^{2i}, i = 1, \ldots, n-1, F_n = x_1 \cdots x_n.$$

Thus,

$$F_i|_{t_\beta} = F_i(-x_2, x_2, \ldots, x_n), i < n, F_n|_{t_\beta} = x_2^2 x_3 \cdots x_n.$$

Since $F_i(-x_2, x_2, \ldots, x_n)$ for $i < n$ obviously constitute a regular sequence, and the number of variables is $n-1$, necessarily $F_n|_{t_\beta} \in (F_1|_{t_\beta}, \ldots, F_{n-1}|_{t_\beta})$. The formality property follows.

Case 4 (A_n). Here the standard coordinates in t satisfy the equality

$$x_1 + \cdots + x_{n+1} = 0.$$

In these coordinates $\beta = x_1 - x_{l+1}$. One can choose the generating invariant polynomials in the form

$$F_i(x_1, \ldots, x_{n+1}) = \sum_{k=1}^{n+1} x_k^i, i = 2, \ldots, n+1.$$

The restrictions have the form

$$F_i(x_1, \ldots, x_n, x_1), i = 2, \ldots, n+1.$$

These polynomials form a regular sequence for $i = 2, \ldots, n$, as required. The proof is complete.

$\square$

3.4. Completion of Proof of Theorem 1

The proof of Theorem 1 follows from Theorem 4 and Proposition 1.

4. Application: Formality of 3-Sasakian Homogeneous Manifolds of Classical Type

4.1. Quaternionic-Kaehler Symmetric Spaces (Wolf Spaces)

In this subsection we present a version of Theorem 3 in terms of the root systems (see Theorems 6 and 7). Let $\mathfrak{g}$ be a compact simple Lie algebra and t be its maximal abelian subalgebra. Consider the complexifications $\mathfrak{g}^c$ and t^c. Thus, t^c is a Cartan subalgebra of $\mathfrak{g}^c$. Let $\Delta = \Delta(\mathfrak{g}^c, t^c)$ denote the root system determined by t^c. Choose the maximal root $\beta \in \Delta$ with respect to some fixed ordering of Δ. As usual, $\mathfrak{g}_\alpha$ denotes the root space of $\alpha \in \Delta$. Define

$$\mathfrak{l}_1 = \{H \in t \mid \beta(H) = 0\} + \sum_{\alpha > 0, \langle \alpha, \beta \rangle = 0} \mathfrak{g} \cap (\mathfrak{g}_\alpha + \mathfrak{g}_{-\alpha}). \tag{1}$$

Put

$$\mathfrak{a}_1 = \mathfrak{g} \cap (\{H_\beta\} + \mathfrak{g}_\beta + \mathfrak{g}_{-\beta}), \tag{2}$$

and

$$\mathfrak{k} = \mathfrak{l}_1 + \mathfrak{a}_1. \tag{3}$$

Theorem 6 (Wolf, [16]). *If G/K is a quaternionic-Kaehler symmetric space, then $K = L_1 \cdot A_1$, where the Lie algebras $\mathfrak{l}_1$ and $\mathfrak{a}_1$ are determined by Equations (1)–(3).*

Theorem 7 ([9], Section 2). *Let $G/K = G/L_1 \cdot A_1$ be the quaternionic symmetric space. Then the homogeneous space G/L_1 is 3-Sasakian. All compact homogeneous Sasakian manifolds are obtained in this way.*

Remark 1. Theorem 7 follows from the description of 3-Sasakian manifolds in [9] together with Theorem 6.

4.2. Proof of Theorem 2

By Theorem 7, any compact homogeneous 3-Sasakian manifold G/H has the form G/L_1 with L_1 given by Theorem 6. One can easily notice that the maximal torus T_{L_1} in L_1 has the Lie algebra of the form $\mathfrak{t}_\beta = \ker \beta$ for the maximal root β. By Theorem 1 the formality property of G/L_1 follows.

5. Conclusions

We have proved that if G is a classical compact Lie group, then the quotient of G by a torus determined by a maximal root, is formal. This result may have important applications in geometry of homogeneous spaces. As an example of such application we present a direct short proof of a result of Fernández, Muñoz and Sanchez about the formality property of some homogeneous 3-Sasakian manifolds.

Funding: This research was funded by National Science Center, Poland, grant No. 2018/31/B/ST1/00053. The APC was funded by National Science Center, Poland.

Acknowledgments: I thank Marisa Fernández and Vicente Muñoz for discussions related to their article [9].

Conflicts of Interest: The authors declare no conflict of interest. The funders had no role in the design of the study; in the collection, analyses, or interpretation of data; in the writing of the manuscript, or in the decision to publish the results.

References

1. Deligne, P.; Griffiths, P.; Morgan, J.; Sullivan, D. Real homotopy theory of Kaehler manifolds. *Invent. Math.* **1975**, *29*, 245–274. [CrossRef]
2. Félix, Y.; Oprea, J.; Tanré, D. *Algebraic Models in Geometry*; Oxford University Press: Oxford, UK, 2008.
3. Tralle, A.; Oprea, J. *Symplectic Manifolds with no Kaehler Structure*; Springer: Berlin, Germany, 1997.
4. Biswas, I.; Fernández, M.; Munoz, V.; Tralle, A. On formality of Sasakian manifolds. *J. Topol.* **2016**, *9*, 161–180. [CrossRef]
5. Amann, M. Non-Formal Homogeneous Spaces. *Math. Z.* **2012**, *274*, 1299–1325. [CrossRef]
6. Morocka-Tralle, I.; Tralle, A. On formality of homogeneous Sasakian manifolds. *Complex Manifolds* **2019**, *6*, 160–169. [CrossRef]
7. Onishchik, A. *Topology of Transitive Transformation Groups*; Johann Ambrosius Barth: Leipzig, Germany, 1993.
8. Stępień, Z. On formality of a class of compact homogeneous spaces. *Geom. Dedic.* **2002**, *93*, 37–45.
9. Fernández, M.; Munoz, V.; Sanchez, J. On $SO(3)$-bundles over the Wolf spaces. *arXiv* **2017**, arXiv:1709.08806.
10. Mehta, M.L. Basic sets of invariant polynomials for finite reflection groups. *Commun. Algebra* **1988**, *16*, 1083–1098. [CrossRef]
11. Greub, V.; Halperin, S.; Vanstone, R. *Curvature, Connections and Cohomology*; Academic Press: New York, NY, USA, 1976; Volume 3.
12. Onishchik, A.; Vinberg, E. *Lie Groups and Lie Algebras III*; Springer: Berlin, Germany, 1994.
13. Boyer, C.; Galicki, K. *Sasakian Geometry*; Oxford University Press: Oxford, UK, 2007.
14. Félix, Y.; Halperin, S.; Thomas, J.-C. *Rational Homotopy Theory*; Springer: Berlin, Germany, 2002.
15. Bourbaki, N. *Groupes et Algebres de Lie*; Hermann: Paris, France, 1968.
16. Wolf, J. Complex homogeneous contact manifolds and quaternionic symmetric spaces. *J. Math. Mech.* **1965**, *14*, 1033–1047.

symmetry

MDPI

Article

Necessary and Sufficient Optimality Conditions for Vector Equilibrium Problems on Hadamard Manifolds

Gabriel Ruiz-Garzón [1,*,†,‡], **Rafaela Osuna-Gómez** [2,‡] **and Jaime Ruiz-Zapatero** [3,‡]

1   Departamento de Estadística e I.O., Universidad de Cádiz, 11405 Cádiz, Spain
2   Departamento de Estadística e I.O., Universidad de Sevilla, 41012 Sevilla, Spain
3   Department of Physics and Astronomy, University College of London, London WC1E 6BT, UK
*   Correspondence: gabriel.ruiz@uca.es
†   Current address: Departamento de Estadística e I.O., Universidad de Cádiz, Campus de Jerez de la Frontera, Avda. de la Universidad s/n, 11405, Jerez de la Frontera, Cádiz, Spain.
‡   These authors contributed equally to this work.

Received: 18 July 2019; Accepted: 8 August 2019; Published: 12 August 2019

Abstract: The aim of this paper is to show the existence and attainability of Karush–Kuhn–Tucker optimality conditions for weakly efficient Pareto points for vector equilibrium problems with the addition of constraints in the novel context of Hadamard manifolds, as opposed to the classical examples of Banach, normed or Hausdorff spaces. More specifically, classical necessary and sufficient conditions for weakly efficient Pareto points to the constrained vector optimization problem are presented. The results described in this article generalize results obtained by Gong (2008) and Wei and Gong (2010) and Feng and Qiu (2014) from Hausdorff topological vector spaces, real normed spaces, and real Banach spaces to Hadamard manifolds, respectively. This is done using a notion of Riemannian symmetric spaces of a noncompact type as special Hadarmard manifolds.

Keywords: vector equilibrium problem; generalized convexity; hadamard manifolds; weakly efficient pareto points

1. Introduction

The pursuit of equilibrium is a ubiquitous horizon in practically all areas of human activity. For example, in economics, the dynamics of offer and demand are typically described as equilibrium problems. In the same way, physical or social phenomena such as the distribution of particles in a container, traffic flow or telecommunication networks can be accurately conceptualized in terms of equilibrium.

However, it was not until Fan [1] that equilibrium theory was applied in the context of Euclidean spaces. Mathematically, the simplest definition of a equilibrium problem consists in finding $x \in S$ such that

$$F(x,y) \geq 0, \forall y \in S$$

where $S \subseteq \mathbb{R}^p$ is a nonempty closed set and $F : \mathbb{R}^p \times \mathbb{R}^p \to \mathbb{R}$ is an equilibrium bifunction, i.e., $F(x,x) = 0$ for all $x \in S$.

Some of the main mathematical problems that can be phrased as equilibrium problems are:

- The weak minimum point of a multiobjective function $f = (f_1, \ldots, f_p)$ over a closed set $S \subseteq \mathbb{R}^p$ is any $\bar{x} \in S$ such that for any $y \in S$, $\exists i$ such that $f_i(y) - f_i(\bar{x}) \geq 0$. Finding a weak minimum point can be reduced to solving an equilibrium problem by virtue of setting

$$F(x,y) = \max_{i=1,\ldots,p} [f_i(y) - f_i(x)].$$

- The Stampacchia variational inequality problem demands finding $\bar{x} \in S$ such that

$$< G(\bar{x}), y - \bar{x} > \geq 0, \ \forall y \in S$$

  where $G : \mathbb{R}^p \to \mathbb{R}^p$ and $S \subseteq \mathbb{R}^p$ is a closed set. This problem is also an equilibrium problem where

$$F(x,y) = < G(x), y - x > .$$

- Nash equilibrium problems in a non-cooperative game with p players where each player i has a set of possible strategies $K_i \subseteq \mathbb{R}^{n_i}$ aim to minimize a loss function $f_i : K \to \mathbb{R}$ with $K = K_1 \times \ldots \times K_p$. Thus, a Nash equilibrium point is any $\bar{x} \in K$ such that no player can reduce its loss by unilaterally changing their strategy, i.e., any $\bar{x} \in K$ such that

$$f_i(\bar{x}) \leq f_i(\bar{x}(y_i))$$

  holds for any $y_i \in K_i$ for any $i = 1, \ldots, p$, with $\bar{x}(y_i))$ denoting the vector obtained from $\bar{x}$ by replacing $\bar{x}_i$ with y_i. Therefore, this problem amounts to solving an equilibrium problem with

$$F(x,y) = \sum_{i=1}^{p} [f_i(x(y_i)) - f_i(x)].$$

Despite their apparent diversity, all the above-mentioned problems can be framed as particular cases of the vector equilibrium problem and thus can all be encompassed in a single mathematical picture. Due to the power of this formulation, it is of great interest to obtain and study the Karush–Kuhn–Tucker (KKT) optimality conditions for the solution of such, more general problems.

Thanks to their capacity to provide such a fundamental insight, vector equilibrium problems are an active branch of non-linear analysis with plenty of publications being made up to this date. For example, in 2003, authors such as Iusem and Sosa [2] studied the relation between equilibrium problems and some auxiliary convex problems. In addition, over the past century, the field of physics departed from euclidean geometry as a space in which to allocate its theories, opting instead for more complex spaces also known as manifolds. A historical landmark that illustrates this example is Einstein's theory of gravity that revolves around the concept of space-time curvature on a Riemannian manifold. Other less known but equally fundamental applications in the fields of physics involve the appearance of symplectic manifolds in the treatment of Hamiltonian vector fields or Noether's theorem.

Smooth Riemannian manifolds are spaces that contain curvature, as opposed to Euclidean spaces which are flat everywhere. This can be mathematically expressed as $ax + by \notin M$, $\forall x, y \in M$, $a, b \in \mathbb{R}$, where M is a Riemannian manifold. Nonetheless, Riemannian geometry constitutes a generalization of the Euclidean case. This can be easily understood by introducing the notion of tangent planes. For any point of a smooth curved space, say a 2-Sphere, it is always possible to define a flat tangent plane to that point; i.e., a Euclidean space. We can think of this in the same way we think of the Earth to be flat at local scales while overall being spherical. Indeed, all curved manifolds locally resemble Euclidean space, which is a vital property for our understanding of them. However, cartography can empirically tell us that flat projections of curved surfaces onto planes fails to faithfully represent the real dimensions of the objects that live on the original curved surface especially at large scales where the locality condition starts weakening. Thus, metricity is no longer trivial and measurements of distances need to account for such curvature.

At this point, we can already see how Euclidean spaces are simply Riemannian manifolds for which the tangent plane to any of its points is identical to the plane itself. Thus, in Euclidean spaces, vectors living of the surface are equivalent to vectors living on its tangent space. It is this key feature of Euclidean geometry that allows for the simple definition of distance as the dot product. Thus, given a vector u, if allocated in an Euclidean space, its length is given by $|u|^2 = < u, u >$. On the other hand, in non-flat spaces it is necessary to account for the distortion of the distances when projected to the tangent space. Riemannian manifolds are those equipped with a so called "metric tensor"; commonly denoted k_{ij}, that allows us to adequately define distances; i.e., $|u|^2 = k_{ij}u^i u^j$. (see Section 2 for more details).

This new definition of length has direct short comings in minimization and equilibrium. The Euclidean line element, the shortest connection between two points on a flat surface, is replaced on manifolds by a geodesic equation which plays the role of straight lines in non-flat spaces. This can be seen from the fact that geodesic curves are solutions to the Euler–Lagrange equations which minimize the functional of the Lagrangian given by the metric of such space, $\mathcal{L} = k_{ij}dx^i dx^j$, and as such describe the trajectories that minimize the action necessary to move from A to B. For example, the orbits of planets obey geodesics despite clearly not being straight in a Euclidean sense.

A Hadamard manifold is a simply connected complete Riemannian manifold of non-positive sectional curvature. The motivation of the study of Hadamard spaces is that they share some properties with Euclidean spaces. One of them is the separation theorem (see Ferreira and Oliveira [3]).

In addition, for any two points in M, there exists a minimal geodesic joining these two points. In a Hadamard manifold, the geodesic between any two points is unique and the exponential map at each point of M is a global diffeomorphism. Moreover, the *exp* map is defined on the whole tangent space ([4]).

However, the minimization of functions on a Hadamard manifold is locally equivalent to the smoothly constrained optimization problem on a Euclidean space, due to the fact that every C^∞ Hadamard manifold can be isometrically embedded in an Euclidean space by virtue of John Nash's embedding theorem. This is consistent with the intuition we previously laid out.

The study of optimization problems on Hadamard manifolds is a powerful tool. This is due to the fact that, generally, solving nonconvex constrained problems in $\mathbb{R}^n$ with the Euclidean metric can be also framed as solving the unconstrained convex minimization problem in the Hadamard manifold feasible set with the affine metric (see [5]). In Colao et al. [5] the existence of solutions for equilibrium problems under some suitable conditions on Hadamard manifolds and their applications to Nash equilibrium for non-cooperative games was studied. In the same way, in Németh [6] the existence and uniqueness results for variational inequality problems on Hadamard manifolds were obtained.

Moreover, many optimization problems cannot be solved in linear spaces, for example, controlled thermonuclear fusion research (see [7]), signal processing, numerical analysis and computer vision (see [8,9]) require Hadamard manifold structures for their modeling. Also, geometrical structures hidden in data sets of machine learning problems are studied in terms of manifolds. In the field of medicine, Hadamard manifolds have been used in the analysis of magnetic resonances to quantify the growth of tumors and consequently deduce their state of progression, as shown by Fletcher et al. [10]. The geometry necessary to understand and perform these techniques is best understood through the use of manifolds and symmetric structures. For example, the set of symmetric positive definite matrices used in magnetic resonance imaging to study Alzheimer's disease [11] is one case in which this translation to manifolds is necessary. In addition, other problems in computer vision, signal processing or learning algorithms employ geodesic curves when addressing optimization problems. Finally, in economics, the search of Nash–Stampacchia equilibria points using Hadamard manifolds has been used by Kristály [12].

It is known that a convex environment has good properties for the search of optimal points. In Ferreira [13], the author gives necessary and sufficient conditions for convex functions on Hadamard manifolds. A significant generalization of the convex functions are the invex functions, introduced by Hanson [14], where the *x-y* vector is replaced by any function $\eta(x, y)$. The main result of invex functions states that a scalar function is invex if and only if every critical point is a global minimum solution. This property is essential to obtain optimal points through algorithms, due to the coincidence of critical points and solutions being always assured. In Barani and Pouryayeli [15] and Hosseini and Pouryayevali [16], the relation between invexity and monotonicity using the mean value theorem is studied. Ruiz-Garzón et al. [17] showed that invexity can be characterized in the context of Riemannian manifolds for both scalar and vector cases, in a similar way to Euclidean spaces. Recently, in Ahmad et al. [18] the authors introduced the log-preinvex and log-invex functions on Riemannian manifolds and the mean value theorem on Cartan-Hadamard manifolds.

In the same way, several authors have studied vector equilibrium problems. Ansari and Flores-Bazán [19] were capable of providing a theorem of existence of solutions to vector quasi-equilibrium problems. Furthermore, a characterization for a weakly efficient Pareto point for the vector equilibrium problems with constraints under convexity conditions on real Hausdorff topological vector spaces were presented by Gong [20]. In the following years, scalarization results for the solutions to the vector equilibrium problems were also given by Gong [21]. Later, optimality conditions for weakly efficient Pareto points to vector equilibrium problems with constraints in real normed spaces were investigated by Wei and Gong [22]. Also, sufficient conditions of weakly efficient Pareto points on real Banach spaces for vector equilibrium and vector optimization problems with constraints under generalized invexity were obtained by Feng and Qiu [23].

Motivated by Gong's works mentioned above, our objective will focus on extending the KKT necessary and sufficient conditions for constrained vector equilibrium problems obtained in topological or normed spaces to other environments like the Hadamard manifolds, not present in the literature up to date of publication. Hence, we propose a generalization that extends the linear space definition to Hadamard manifolds, by virtue of substituting line segments by geodesic arcs. We will see that the KKT classic conditions for constrained vector optimization are a particular case of the ones obtained for constrained vector equilibrium problem.

The organization of the paper is as follows: In Section 2, we discuss notation, differentials and invex function concepts on Hadamard manifolds. Section 3 is devoted to proving the main results obtained in this paper, and studying the necessary and sufficient optimality conditions for weakly efficient points of the constrained vector equilibrium problem. Section 4 dwells on how the previous results can be reduced to classical KKT conditions for constrained vector optimization problems, first obtained by William Karush [24] and rediscovered by Harold Kuhn and Albert Tucker [25]. Finally, an example is presented as well as the final conclusions.

2. Preliminaries

Let M be a C^∞-manifold modeled on a Hilbert space H endowed with a Riemannian metric g_x on a tangent space T_xM. We denote by T_xM the tangent space of M at x, by $TM = \bigcup_{x \in M} T_xM$ the tangent bundle of M, by $\bar{T}M$ an open neighborhood of the submanifold M of TM. The corresponding norm is denoted by $\|.\|_x$ and the length of a piecewise C^1 curve $\alpha : [a, b] \to M$ is defined by

$$L(\alpha) = \int_a^b \|\alpha'(t)\|_{\alpha(t)} dt.$$

We define d as the distance which induces the original topology on M such that

$$d(x, y) = \inf\{L(\alpha)| \ \alpha \text{ is a piecewise } C^1 \text{ curve joining } x \text{ and } y \ \forall x, y \in M\}.$$

If d is the distance induced by the Riemannian metric k_{ij} then any Riemannian manifold (M, k_{ij}) can be converted into a metric space (M, d). The derivatives of the curves at a point x on the manifold lies in a vector space T_xM. Whatever path α joining x and y in M such that $L(\alpha) = d(x, y)$ is a geodesic.

Let $\exp : \bar{T}M \to M$ be the Riemannian exponential map defined as $\exp_x(V) = \alpha_V(1)$ for every $V \in \bar{T}M$, where α_V is the geodesic starting at x with velocity V (i.e., $\alpha(0) = x$, $\alpha'(0) = V$).

Assume now that η is a map $\eta : M \times M \to TM$ defined on the product manifold such that

$$\eta(x, y) \in T_yM, \ \forall x, y \in M.$$

Definition 1. *[26] A subset S_1 of M is considered totally convex if S_1 contains every geodesic $\alpha_{x,y}$ of M whose endpoints x and y belong to S_1.*

On a Hadamard manifold M, we can define the function η as $\eta(x,y) = \alpha'_{x,y}(0)$ for all $x, y \in M$. This function plays the same role of $x - y \in \mathbb{R}^n$. Here $\alpha_{x,y}$ is the unique minimal geodesic joining y to x as follows

$$\alpha_{x,y} = \exp_y(\lambda \exp_y^{-1} x) \quad \forall \lambda \in [0,1].$$

Example 1. *Let $M = \mathbb{R}_{++} = \{y \in \mathbb{R} : y > 0\}$ endowed with the Riemannian metric defined by $g(y) = y^{-2}$ be a Hadamard manifold. Hyperbolic spaces and geodesic spaces, more precisely, a Busemann non-positive curvature (NPC) space are examples of Hadarmard manifolds.*

We will need an adequate concept of the differential:

Definition 2. *[27] A mapping $f_i : M \to \mathbb{R}$ is said to be a differential map along the geodesic $\alpha_{x,y}$ at $y \in M$ if and only if the limit*

$$f'_i(y) = \lim_{\lambda \to 0} \frac{f_i(\exp_y(\lambda \eta(x,y))) - f_i(y)}{\lambda \|\eta(x,y)\|}$$

exists.

The gradient of a real-valued C^∞ function $f = (f_1, \ldots f_p) : S_1 \subseteq M \to \mathbb{R}^n$ on M in x, denoted by $grad f_x = (f'_1(x), f'_2(x), \ldots, f'_n(x))$, is the unique vector in $T_x M$ such that $df_x(X) = \langle grad f_x, X \rangle$ for all X in $T_x M$ is the differential of f at $\bar{x}$ of X.

Remark 1. *The differential of f at $\bar{x}$ of X is similar to the definition of directional derivative in the Euclidean space.*

Let $S_1 \subset M$ be a nonempty open totally convex subset and let $F : S_1 \times S_1 \to \mathbb{R}^p$, $g : S_1 \to \mathbb{R}^p$ be mappings.

Definition 3. *We define the constraint set $S = \{x \in S_1 : g(x) \in -\mathbb{R}^p_+\}$ and consider the vector equilibrium problem with constraints (VEPC): find $x \in S$ such that*

$$F(x,y) \notin -\mathbb{R}^p_+ \setminus \{0\}, \forall y \in S$$

where $\mathbb{R}^p_+$ is the non-negative orthant of $\mathbb{R}^p$.

We recall the classical concept:

Definition 4. *A vector $x \in S$ satisfying $F(x,y) \notin -int\, \mathbb{R}^p_+$, $\forall y \in S$ is called a weakly efficient Pareto point to the VEPC.*

Notation 1. *We denote as $H_x(y) = F(x,y)$, $\forall y \in S_1$, given $x \in S$, where $H : S_1 \to \mathbb{R}^p$ is a mapping.*

Inspired by the concept of convexity on a linear space, the notion of invexity function concept on Hadamard manifolds has become a successful tool in vector optimization. This generalized definition was notably provided by Hanson in [14].

Definition 5. *Let S_1 be a nonempty open totally convex subset of a Hadamard manifold M. A differentiable $h : S_1 \to \mathbb{R}^p$ function is said to be a $\mathbb{R}^p_+$-invex at $\bar{x} \in S_1$ respect to $\eta : M \times M \to TM$ if there exist $\eta(x, \bar{x}) \in T_{\bar{x}} M$ such that*

$$h(x) - h(\bar{x}) - dh_{\bar{x}}(\eta(x, \bar{x})) \in \mathbb{R}^p_+.$$

Using the previously stated definitions, we can obtain the sufficient conditions for optimality by virtue of the assumption of invexity of the functions of the problem.

3. Main Results

Next, we will obtain a characterization for the weakly efficient points of VEPC through the application of necessary and sufficient optimality conditions. We start with the necessary conditions:

Theorem 1. *[Necessary KKT-conditions] Let S_1 be a nonempty open totally convex subset of a Hadamard manifold M and let $F : S_1 \times S_1 \to \mathbb{R}^p$, $g : S_1 \to \mathbb{R}^p$, $\eta : M \times M \to TM$ be mappings. Let $F(\bar{x}, \bar{x}) = H_{\bar{x}}(\bar{x}) = 0$. Assume that H and g are differentiable at $\bar{x} \in S$. Furthermore, assume that there exists $x_1 \in S_1$ such that $g(\bar{x}) + dg_{\bar{x}}(\eta(x_1, \bar{x})) \in -int\,\mathbb{R}^p_+$. If $\bar{x}$ is a weakly efficient Pareto point to the VEPC, then there exists $v \in \mathbb{R}^p_+ \setminus \{0\}$, $u \in \mathbb{R}^p_+$ such that*

$$v\,dH_{\bar{x}}(\eta(x, \bar{x})) + u\,dg_{\bar{x}}(\eta(x, \bar{x})) \geq 0, \ \forall x \in S_1 \tag{1}$$

$$u\,g(\bar{x}) = 0. \tag{2}$$

Proof. Let there be $\bar{x} \in S$ as a weakly efficient Pareto point to the VEPC. We denote by

$$W = \{(y, z) \in \mathbb{R}^p \times \mathbb{R}^p : \text{there exists} \quad x \in S_1, \quad \text{such that} \quad y - dH_{\bar{x}}(\eta(x, \bar{x})) \in int\,\mathbb{R}^p_+,$$

$$z - [g(\bar{x}) + dg_{\bar{x}}(\eta(x, \bar{x}))] \in int\,\mathbb{R}^p_+\}.$$

It may be noted that W is a nonempty open totally convex set. This proof can be divided into five steps:

Step 1. We have to prove that $(0, 0) \notin W$. By reduction ad absurdum, if $(0, 0) \in W \Rightarrow \exists x_0 \in S_1$, such that

$$dH_{\bar{x}}(\eta(x_0, \bar{x})) \in -int\,\mathbb{R}^p_+, \quad g(\bar{x}) + dg_{\bar{x}}(\eta(x_0, \bar{x})) \in -int\,\mathbb{R}^p_+. \tag{3}$$

From the differentiability we obtain that

$$dH_{\bar{x}}(\eta(x_0, \bar{x})) = \lim_{\lambda \to 0} \frac{1}{\lambda}[H_{\bar{x}}(exp_{\bar{x}}(\lambda\eta(x_0, \bar{x})) - H_{\bar{x}}(\bar{x})] \in -int\,\mathbb{R}^p_+ \tag{4}$$

$$g(\bar{x}) + dg_{\bar{x}}(\eta(x_0, \bar{x})) = g(\bar{x}) + \lim_{\lambda \to 0} \frac{1}{\lambda}[g(exp_{\bar{x}}(\lambda\eta(x_0, \bar{x}))) - g(\bar{x})] \in -int\,\mathbb{R}^p_+. \tag{5}$$

As $-int\,\mathbb{R}^p_+$ is an open set, then $\exists \lambda_0$, $0 < \lambda_0 < 1$ such that

$$\frac{1}{\lambda_0}[H_{\bar{x}}(exp_{\bar{x}}(\lambda_0\eta(x_0, \bar{x}))) - H_{\bar{x}}(\bar{x})] \in -int\,\mathbb{R}^p_+ \tag{6}$$

$$g(\bar{x}) + \frac{1}{\lambda_0}[g(exp_{\bar{x}}(\lambda_0\eta(x_0, \bar{x}))) - g(\bar{x})] \in -int\,\mathbb{R}^p_+. \tag{7}$$

By hypothesis, from $g(\bar{x}) \in -\mathbb{R}^p_+$, $F(\bar{x}, \bar{x}) = H_{\bar{x}}(\bar{x}) = 0$, and $\frac{1}{\lambda_0} > 1$, then

$$H_{\bar{x}}[exp_{\bar{x}}(\lambda_0\eta(x_0, \bar{x}))] \in -int\,\mathbb{R}^p_+ \quad \text{and} \quad g(exp_{\bar{x}}(\lambda_0\eta(x_0, \bar{x}))) \in -int\,\mathbb{R}^p_+. \tag{8}$$

As S_1 is a totally convex set we have that

$$exp_{\bar{x}}(\lambda_0\eta(x_0, \bar{x})) \in S_1, \quad F(\bar{x}, exp_{\bar{x}}(\lambda_0\eta(x_0, \bar{x}))) \in -int\,\mathbb{R}^p_+ \tag{9}$$

and

$$g(exp_{\bar{x}}(\lambda_0\eta(x_0, \bar{x}))) \in -int\,\mathbb{R}^p_+ \tag{10}$$

stands in contradiction with $\bar{x} \in S$ as a weakly efficient Pareto point to the VEPC, consequently $(0,0) \notin W$.

Step 2. We will prove that there exists a multiplier $v \in \mathbb{R}_+^p$. As W is an open set and the separation theorem holds (see Theorem 2.13 and Remark 2.14 in [28]) or [3]), there exists $(v,u) \neq (0,0) \in \mathbb{R}^p \times \mathbb{R}^p$ such that

$$vy + uz > 0, \; \forall (y,z) \in W. \tag{11}$$

Let $(y,z) \in W$ be a point then $\exists x \in S_1$ such that

$$y - dH_{\bar{x}}(\eta(x,\bar{x})) \in int \, \mathbb{R}_+^p, \quad z - [g(\bar{x}) + dg_{\bar{x}}(\eta(x,\bar{x}))] \in int \, \mathbb{R}_+^p. \tag{12}$$

For any $r \in int \, \mathbb{R}_+^p, s \in int \, \mathbb{R}_+^p, t', t'' > 0$, we have $(y + t'r, z) \in W$ and $(y, z + t''s) \in W$. From Equation (11) we have that

$$v(y + t'r) + u(z) > 0, \; \forall r \in int \, \mathbb{R}_+^p, \; t' > 0. \tag{13}$$

Then

$$vr > \frac{-uz - vy}{t'}. \tag{14}$$

Letting $t' \to \infty$ we get $vr \geq 0$, $\forall r \in int \, \mathbb{R}_+^p$ and therefore $vr \geq 0$ for all $r \in \mathbb{R}_+^p$, that is $v \in \mathbb{R}_+^p$. In the same way, we can show that $u \in \mathbb{R}_+^p$.

Step 3. We will prove that $v \neq 0$, thus is, $v \in \mathbb{R}_+^p \setminus \{0\}$. By reduction ad absurdum, if $v = 0$, from Equation (11) we get

$$uz > 0, \; \forall (y,z) \in W. \tag{15}$$

According to the hypothesis, $\exists x_1 \in S_1$ such that $g(\bar{x}) + dg_{\bar{x}}(\eta(x_1,\bar{x})) \in -int \, \mathbb{R}_+^p$; then, we obtain

$$(dH_{\bar{x}}(\eta(x_1,\bar{x})) + r, g(\bar{x}) + dg_{\bar{x}}(\eta(x_1,\bar{x})) + s) \in W, \; \forall r \in int \, \mathbb{R}_+^p \quad \forall s \in int \, \mathbb{R}_+^p. \tag{16}$$

Therefore, from Equation (11) we have that

$$u[g(\bar{x}) + dg_{\bar{x}}(\eta(x_1,\bar{x})) + s] > 0, \quad \forall s \in int \, \mathbb{R}_+^p \tag{17}$$

$$us > -u[g(\bar{x}) + dg_{\bar{x}}(\eta(x_1,\bar{x}))]. \tag{18}$$

As $[g(\bar{x}) + dg_{\bar{x}}(\eta(x_1,\bar{x}))] \in -int \, \mathbb{R}_+^p$, and if $s = 0$, we get $u \cdot 0 = 0 > 0$, which implies a contradiction, thus $v \neq 0$.

Step 4. We will prove the first KKT condition. Since

$$(dH_{\bar{x}}(\eta(x,\bar{x})) + r, g(\bar{x}) + dg_{\bar{x}}(\eta(x,\bar{x})) + s) \in W, \; x \in S_1, \; r \in int \, \mathbb{R}_+^p, \; s \in int \, \mathbb{R}_+^p. \tag{19}$$

From Equation (11) we get

$$v[dH_{\bar{x}}(\eta(x,\bar{x})) + r] + u[g(\bar{x}) + dg_{\bar{x}}(\eta(x,\bar{x})) + s] > 0, \; \forall x \in S_1, \; r \in int \, \mathbb{R}_+^p, \; s \in int \, \mathbb{R}_+^p. \tag{20}$$

Letting $r \to 0, s \to 0$, we obtain

$$vdH_{\bar{x}}(\eta(x,\bar{x})) + u[g(\bar{x}) + dg_{\bar{x}}(\eta(x,\bar{x}))] \geq 0, \quad \forall x \in S_1. \tag{21}$$

Step 5. We will prove the second KKT condition. As

$$(dH_{\bar{x}}(\eta(\bar{x},\bar{x})) + t'r, g(\bar{x}) + dg_{\bar{x}}(\eta(\bar{x},\bar{x})) + t's) \in W, \; \forall r \in int \, \mathbb{R}_+^p, \; s \in int \, \mathbb{R}_+^p, \; t' > 0. \tag{22}$$

From Equation (11) we have that

$$v[dH_{\bar{x}}(\eta(\bar{x}, \bar{x})) + t'r] + u[g(\bar{x}) + dg_{\bar{x}}(\eta(\bar{x}, \bar{x})) + t's] = t'vr + ug(\bar{x}) + t'us > 0. \qquad (23)$$

Letting $t' \to 0$, we obtain $ug(\bar{x}) \geq 0$. Noting that $g(\bar{x}) \in -\mathbb{R}^p_+$ and $u \in \mathbb{R}^p_+$, we have that $ug(\bar{x}) \leq 0$, in consequence

$$ug(\bar{x}) = 0 \qquad (24)$$

and therefore $\exists v \in \mathbb{R}^p_+ \setminus \{0\}, u \in \mathbb{R}^p_+$ such that KKT conditions

$$vdH_{\bar{x}}(\eta(x, \bar{x})) + udg_{\bar{x}}(\eta(x, \bar{x})) \geq 0, \forall x \in S_1 \qquad (25)$$

$$ug(\bar{x}) = 0 \qquad (26)$$

hold. $\square$

Let us see now the reciprocal of the previous theorem. To obtain it we first need conditions of invexity.

Theorem 2. *[Sufficient KKT-conditions] Let S_1 be a nonempty open totally convex subset of Hadamard manifold M and let $F : S_1 \times S_1 \to \mathbb{R}^p$, $g : S_1 \to \mathbb{R}^p$ be mappings. Let $F(\bar{x}, \bar{x}) = H(\bar{x}) = 0$. Assume that H and g are differentiable at $\bar{x} \in S$. H and g are $\mathbb{R}^p_+$-invex at $\bar{x}$ respect to η on S_1. If there exist $v \in \mathbb{R}^p_+ \setminus \{0\}$ and $u \in \mathbb{R}^p_+$ such that*

$$vdH_{\bar{x}}(\eta(x, \bar{x})) + udg_{\bar{x}}(\eta(x, \bar{x})) \geq 0, \forall x \in S_1 \qquad (27)$$

$$ug(\bar{x}) = 0 \qquad (28)$$

then $\bar{x}$ is a weakly efficient Pareto point to the VEPC.

Proof. On the assumption that H and g are $\mathbb{R}^p_+$-invex at $\bar{x}$ respect to η on S_1 then

$$dH_{\bar{x}}(\eta(x, \bar{x})) \in H_{\bar{x}}(x) - H_{\bar{x}}(\bar{x}) - \mathbb{R}^p_+ = H_{\bar{x}}(x) - \mathbb{R}^p_+, \forall x \in S_1 \qquad (29)$$

$$dg_{\bar{x}}(\eta(x, \bar{x})) \in g(x) - g(\bar{x}) - \mathbb{R}^p_+, \forall x \in S_1. \qquad (30)$$

From $v \in \mathbb{R}^p_+ \setminus \{0\}, u \in \mathbb{R}^p_+$ and (27) we obtain that

$$vH_{\bar{x}}(x) + u(g(x) - g(\bar{x})) = vdH_{\bar{x}}(\eta(x, \bar{x})) + udg_{\bar{x}}(\eta(x, \bar{x})) \geq 0, \quad \forall x \in S_1. \qquad (31)$$

From hypothesis (28), we get on the one hand that:

$$vH_{\bar{x}}(x) + ug(x) \geq 0, \quad \forall x \in S_1. \qquad (32)$$

On the other hand, we will show that $\bar{x}$ is a weakly efficient Pareto point to the VEPC. If not, consequently by definition $\exists y_0 \in S$ such that

$$F(\bar{x}, y_0) \in -int\,\mathbb{R}^p_+. \qquad (33)$$

From $v \in \mathbb{R}^p_+ \setminus \{0\} \Rightarrow vF(\bar{x}, y_0) < 0$.
Since $y_0 \in S$, we have $g(y_0) \in -\mathbb{R}^p_+$, so $ug(y_0) \leq 0$ because of $u \in \mathbb{R}^p_+$ and then

$$vF(\bar{x}, y_0) + ug(y_0) < 0 \qquad (34)$$

stands in contradiction with (32) and therefore $\bar{x}$ is a weakly efficient Pareto point to the VEPC. $\square$

Remark 2. *Theorem 3.1 in [20] on real Hausdorff topological vector spaces and Theorem 3.2 and Theorem 3.4 in [22] on real normed spaces are particular cases of Theorems 1 and 2 obtained in this paper on Hadamard manifolds. The same is true for Theorems 3.1 and 3.3 in [23] on real Banach spaces.*

To sum up, we obtain the KKT optimality conditions for weakly efficient Pareto points to the vector equilibrium problems with constraints. These results are not only necessary but also sufficient.

4. Application

As a particular case of the results obtained in the previous section, we will obtain the optimality conditions of KKT for constrained vector optimization problems.

Let us consider the constrained multiobjective programming (CVOP) defined as:

$$(CVOP) \quad \min f(x)$$
$$\text{subject to:}$$
$$g(x) \leq 0$$
$$x \in X \subseteq M$$

where $f = (f_1, \dots f_p) : X \subseteq M \to \mathbb{R}^p$, $g = (g_1, \dots, g_m) : X \subseteq M \to \mathbb{R}^m$ are differentiable multiobjective functions on the open set $X \subseteq M$ and let M be a Hadamard manifold.

As a consequence of the previous theorems and considering CVOP as a particular case of VEPC we have the KKT classical conditions.

Corollary 1. *Let S_1 be a nonempty open totally convex subset of Hadamard manifold M and let $f, g : S_1 \to \mathbb{R}^p$ be mappings. Assume that f and g are differentiable at $\bar{x} \in S$. Furthermore, assume that there exists $x_1 \in S_1$ such that $g(\bar{x}) + dg_{\bar{x}}(\eta(x_1, \bar{x})) \in -int\,\mathbb{R}^p_+$. If $\bar{x}$ is a weakly efficient Pareto point to the CVOP, then there exist $v \in \mathbb{R}^p_+ \setminus \{0\}$, $u \in \mathbb{R}^p_+$ such that*

$$v\,df_{\bar{x}}(\eta(x, \bar{x})) + u\,dg_{\bar{x}}(\eta(x, \bar{x})) \geq 0, \forall x \in S_1 \tag{35}$$

$$ug(\bar{x}) = 0. \tag{36}$$

Corollary 2. *Let S_1 be a nonempty open totally convex subset of Hadamard manifold M and let $f, g : S_1 \to \mathbb{R}^p$ be mappings. Assume that f and g are differentiable at $\bar{x} \in S$. Assume that f and g are differentiable at $\bar{x} \in S$ and f and g are $\mathbb{R}^p_+$-invex respect at $\bar{x}$ to η on S_1. If there exist $v \in \mathbb{R}^p_+ \setminus \{0\}$, $u \in \mathbb{R}^p_+$ such that*

$$v\,df_{\bar{x}}(\eta(x, \bar{x})) + u\,dg_{\bar{x}}(\eta(x, \bar{x})) \geq 0, \forall x \in S_1 \tag{37}$$

$$ug(\bar{x}) = 0 \tag{38}$$

then $\bar{x}$ is a weakly efficient Pareto point to the CVOP.

Proof. The proofs are similar to those already shown without further considering CVOP as particular cases of VEPC just by taking $F(x, y) = \max_{i=1,\dots,p}[f_i(y) - f_i(x)], \forall x, y \in M$. □

Remark 3. *Theorem 4.4 in [20] on real Hausdorff topological vector spaces and Corollary 3.3 in [23] on real Banach spaces are particular cases of Corollaries 1 and 2 obtained in this paper on Hadamard manifolds. Moreover, these results also coincide with Corollary 3.8 given by Ruiz-Garzón et al. [17].*

We illustrate the previous results with another example:

Example 2. *Let us consider the set* $\Omega = \{p = (p_1, p_2) \in \mathbb{R}^2 : p_2 > 0\}$. *Let K be a 2x2 matrix defined by* $K(p) = (k_{ij}(p))$ *with*

$$k_{11}(p) = k_{22}(p) = \frac{1}{p_2^2}, \quad k_{12}(p) = k_{21}(p) = 0.$$

Endowing Ω *with the Riemannian metric* $\ll u, v \gg = < K(p)u, v >$, *we obtain a complete Riemannian manifold* $\mathbb{H}^2$, *namely, the upper half-plane model of a hyperbolic space and* grad $f(p) = K(p)^{-1}\nabla f(p)$.
Consider the CVOP:

$$\text{(CVOP)} \quad Min f(p) = (f_1, f_2)(p) = (p_1, \ln p_2)$$
$$\text{subject to:}$$

$$g_1(p) = 2p_1 - 2 \geq 0$$

$$g_2(p) = p_2 - 1 \geq 0$$

Given $\overline{p} = (1, 1)$ *using the Riemannian metric k and* f, g *is* $\mathbb{R}_+^2$-*invex at* $\overline{p}$ *respect to* $\eta(p, \overline{p}) = 2p - \overline{p}$ *and there exists* $q = \eta(p, \overline{p}) = (0, 1)$ *we have that*

$$df_{1(\overline{p})}(q) = < \begin{pmatrix} p_2^2 & 0 \\ 0 & p_2^2 \end{pmatrix} \begin{pmatrix} 1 \\ 0 \end{pmatrix}, \begin{pmatrix} 0 \\ 1 \end{pmatrix} >= (p_2^2, 0) \begin{pmatrix} 0 \\ 1 \end{pmatrix} = 0$$

$$df_{2(\overline{p})}(q) = < \begin{pmatrix} p_2^2 & 0 \\ 0 & p_2^2 \end{pmatrix} \begin{pmatrix} 0 \\ p_2^{-1} \end{pmatrix}, \begin{pmatrix} 0 \\ 1 \end{pmatrix} >= (0, p_2) \begin{pmatrix} 0 \\ 1 \end{pmatrix} = p_2.$$

The

$$dg_{1(\overline{p})}(q) = < \begin{pmatrix} p_2^2 & 0 \\ 0 & p_2^2 \end{pmatrix} \begin{pmatrix} 2 \\ 0 \end{pmatrix}, \begin{pmatrix} 0 \\ 1 \end{pmatrix} >= (2p_2^2, 0) \begin{pmatrix} 0 \\ 1 \end{pmatrix} = 0$$

$$dg_{2(\overline{p})}(q) = < \begin{pmatrix} p_2^2 & 0 \\ 0 & p_2^2 \end{pmatrix} \begin{pmatrix} 0 \\ 1 \end{pmatrix}, \begin{pmatrix} 0 \\ 1 \end{pmatrix} >= (0, p_2^2) \begin{pmatrix} 0 \\ 1 \end{pmatrix} = p_2^2.$$

We have

$$df_{\overline{p}}(q) = (df_{1(\overline{p})}(q), df_{2(\overline{p})}(q)) = (0, p_2)$$

$$dg_{\overline{p}}(q) = (dg_{1(\overline{p})}(q), dg_{2(\overline{p})}(q)) = (0, p_2^2)$$

and therefore there exists $v = u = (1, 0)$*such that*

$$vdf_{\hat{x}}(q) + udg_{\hat{x}}(q) = 0$$

$$ug(\bar{x}) = 0$$

and then $\overline{p}$ *is a weakly efficient Pareto point to the CVOP.*

5. Conclusions

In conclusion, we have shown the existence of KKT optimality conditions for weakly efficient Pareto points to the equilibrium vector problems with constraints on Hadamard manifolds, in particular, to constrained vector optimization problems. The main requirement we present for such characterization is the substitution of the segments by geodesics due to the introduction of non-euclidean spaces. This has proven to entail:

- The need for an extension of the concept of convex set to that of totally convex.
- The use of an adequate definition of differential functions in similar terms to those of directional derivatives in Euclidean space using an exponential Riemannian map.

- Generalizing the invexity definition by extending its classical definition given by Hanson [14] in order to obtain sufficient optimality conditions.

Thus, our study provides evidence of the logical continuity of the KKT formulation when extended to other contexts different from Banach spaces or norms, given in the literature by Gong [20] and Wei and Gong [22] and Feng and Qiu [23].

The strength of our method lays on the fact that it allows us to transform non-convex problems in Euclidean spaces to convex problems on Hadamard spaces in which the known properties of convexity can be safely applied. On the other hand, the weakness is that the method requires the manifold to have a nonpositive sectional curvature which limits the cases in which it can be employed. In addition, the dimensions of the Euclidean space tend to be larger than the manifold dimensions, making this approach sometimes not convenient.

The principal contribution of this paper is to obtain the classical KKT optimality conditions for vector equilibrium problems on Hadamard spaces, an unexplored field up to this date.

Finally, it would be interesting to continue studies in this line of research by virtue of considering other types of solutions to the vector equilibrium problem with constraints to which similar generalization have yet not been proposed.

Author Contributions: All authors contributed equally to this research and in writing the paper

Funding: The research in this paper has been partially supported by Ministerio de Economía y Competitividad, Spain, through grant MTM2015-66185-P.

Acknowledgments: We thank the four anonymous referees very much for their many valuable remarks and suggestions that definitely improved the final version of this paper.

Conflicts of Interest: The authors declare no conflict of interest.

References

1. Fan, K. A generalization of Tychonoff's fixed point theorem. *Math. Ann.* **1961**, *142*, 305–310. [CrossRef]
2. Iusem A.N.; Sosa, W. New existence results for equilibrium problems. *Nonlinear Anal.* **2003**, *52*, 621–635. [CrossRef]
3. Ferreira, O.P.; Oliveira, P.R. Proximal point algorithm on Riemannian manifolds. *Optimization* **2002**, *51*, 257–270. [CrossRef]
4. Neeb, K.H. A Cartan-Hadamard Theorem for Banach-Finsler Manifolds. *Geom. Dedicata* **2002**, *95*, 115–156. [CrossRef]
5. Colao, V.; López, G.; Marino, G.; Martín-Márquez, V. Equilibrium problems in Hadamard manifolds. *J. Math. Anal. Appl.* **2012**, *38*, 61–77. [CrossRef]
6. Németh, S.Z. Variational inequalities on Hadamard manifolds. *Nonlinear Anal.* **2003**, *52*, 1491–1498. [CrossRef]
7. Absil, P.A.; Mahony, R.; Sepulchre, R. *Optimization Algorithms on Matrix Manifolds*; Princeton University Press: Princeton, NJ, USA, 2008.
8. Nishimori, Y.; Akaho, S. Learning algorithms utilizing quasigeodesic flows on the Stiefel manifold, *Neurocoputing* **2005**, *67*, 106–135. [CrossRef]
9. Turaga, P.; Veeraraghavan, A.; Chellapa, R. Statistical Analysis on Stiefel and Grasmann manifolds with applications in computer vision. In Proceedings of the IEEE Conference on Computer Vision and Pattern Recognition, Anchorage, AK, USA, 23–28 June 2008.
10. Fletcher, P.T.; Venkatasubramanian, S.; Joshi, S. The geometric median on Riemannian manifolds with application to robust atlas estimation. *NeuroImage* **2009**, *45*, 5143–5152. [CrossRef]
11. Kim, H.J.; Adluru, N.; Bendlin, B.B.; Johnson, S.C.; Vemuri, B.C.; Singh, V. Canonical Correlation Analysis on Riemannian Manifolds and Its Applications. In *Lecture Notes in Computer Science*; Fleet, D., Pajdla, T., Schiele, B., Tuytelaars, T., Eds.; Springer: Berlin, Germany, 2014; Volume 8690, pp. 251–267.
12. Kristály, A. Nash-type equilibria on Riemannian manifolds: A variational approach. *J. Math. Pures Appl.* **2014**, *101*, 660–688. [CrossRef]
13. Ferreira, O.P. Dini derivative and a characterization for Lipschiz and convex functions on Riemannian manifolds. *Nonlinear Anal.* **2008**, *68*, 1517–1528. [CrossRef]
14. Hanson, M.A. On sufficiency of the Khun-Tucker conditions. *J. Math. Anal. Appl.* **1981**, *80*, 545–550. [CrossRef]

15. Barani A.; Pouryayevali, M.R. Invariant monotone vector fields on Riemannian manifolds. *Nonlinear Anal.* **2009**, *70*, 1850–1861. [CrossRef]

16. Hosseini, S.; Pouryayevali, M.R. Generalized gradients and characterization of epi-Lipschitz sets in Riemannian manifolds. *Nonlinear Anal.* **2011**, *74*, 3884–3895. [CrossRef]

17. Ruiz-Garzón, G.; Osuna-Gómez, R.; Rufián-Lizana A.; Hernández-Jiménez, B. Optimality and duality on Riemannian manifolds. *Taiwanese J. Math.***2018**, *22*, 1245–1259. [CrossRef]

18. Ahmad, I.; Khan, M.A.; Ishan, A.A. Generalized Geodesic Convexity on Riemannian manifolds. *Mathematics* **2019**, *7*, 547. [CrossRef]

19. Ansari, Q.H.; Flores-Bazán, F. Generalized vector quasi-equilibrium problems with applications. *J. Math. Anal. Appl.* **2003**, *277*, 246–256. [CrossRef]

20. Gong, X.H. Optimality conditions for vector equilibrium problems. *J. Math. Anal. Appl.* **2008**, *342*, 1455–1466. [CrossRef]

21. Gong, X.H. Scalarization and optimality conditions for vector equilibrium problems. *Nonlinear Anal.* **2010**, *73*, 3598–3612. [CrossRef]

22. Wei, Z.F.; Gong, X.H. Kuhn-Tucker Optimality Conditions for Vector Equilibrium Problems. *J. Inequalities Appl.* **2010**, *2010*, 842715. [CrossRef]

23. Feng, Y.; Qiu, Q. Optimality conditions for vector equilibrium problems with constraint in Banach spaces. *Optim. Lett.* **2014**, *8*, 1391–1944. [CrossRef]

24. Karush, W. Minima of Functions of Several Variables with Inequalities as Side Conditions. Master's Thesis, Department of Mathematics, University of Chicago, Chicago, IL, USA, 1939.

25. Khun, H.W.; Tucker, A.W. Nonlinear Programming. In Proceedings of the Second Berkeley Symposium on Mathematical Statistics and Probability, Statistical Laboratory of the University of California, Berkeley, CA, USA, 31 July–12 August 1950; University of California Press: Berkeley, CA, USA, 1951; pp. 481–492

26. Bangert, V. Totally convex sets in complete Riemannian manifolds. *J. Differ. Geom.* **1981**, *16*, 333–345. [CrossRef]

27. Zhou, L.W.; Huang, N.J. Roughly Geodesic B-invex and Optimization problem on Hadamard Manifolds. *Taiwanese J. Math.* **2013**, *17*, 833–855. [CrossRef]

28. Khajejpour, S.; Pouryayevali, M.R. Convexity of distance function to convex subsets of Riemannian manifolds. *arXiv* **2018**, arXiv:1802.09192.

symmetry

MDPI

Article

Geodesic Chord Property and Hypersurfaces of Space Forms

Dong-Soo Kim [1], Young Ho Kim [2],* and Dae Won Yoon [3]

[1] Department of Mathematics, Chonnam National University, Gwangju 61186, Korea
[2] Department of Mathematics, Kyungpook National University, Daegu 41566, Korea
[3] Department of Mathematics Education and RINS, Gyeongsang National University, Jinju 52828, Korea
* Correspondence: yhkim@knu.ac.kr; Tel.: +82-53-950-7325

Received: 22 July 2019; Accepted: 14 August 2019; Published: 16 August 2019

Abstract: In the Euclidean space $\mathbb{E}^n$, hyperplanes, hyperspheres and hypercylinders are the only isoparametric hypersurfaces. These hypersurfaces are also the only ones with chord property, that is, the chord connecting two points on them meets the hypersurfaces at the same angle at the two points. In this paper, we investigate hypersurfaces in nonflat space forms with the so-called *geodesic chord property* and classify such hypersurfaces completely.

Keywords: geodesic chord property; hypersphere; hyperbolic space; isoparametric hypersurface

1. Introduction

A circle in the plane $\mathbb{E}^2$ is characterized as a closed curve with the *chord property* that the chord connecting any two points on it meets the curve at the same angle at the two end points ([1], pp. 160–162).

For space curves, B.-Y. Chen et al. showed that a W-curve is characterized as a curve in the Euclidean space $\mathbb{E}^n$ with the property that the chord joining any two points on the curve meets the curve at the same angle at the two points, that is, as a curve in the Euclidean space $\mathbb{E}^n$ with the chord property ([2]).

For hypersurfaces in the Euclidean n-space $\mathbb{E}^n$ which satisfies the chord property, D.-S. Kim and Y. H. Kim established the following classification theorem ([3]). See also [4–6].

Proposition 1. *Let us consider a hypersurface M in the Euclidean space $\mathbb{E}^n$. Then the following are equivalent:*

1. *M satisfies the chord property.*
2. *The Gauss map G of M satisfies $G(x) = Ax + b$ for some $n \times n$ matrix A and a vector $b \in \mathbb{E}^n$.*
3. *M is an isoparametric hypersurface.*
4. *M is contained in one of the following hypersurfaces: $\mathbb{E}^{n-1}$, $S^{n-1}(r)$, $S^{p-1}(r) \times \mathbb{E}^{n-p}$.*

In this paper, we consider hypersurfaces in the n-dimensional space form $\bar{M}^n(c)$ with nonzero constant sectional curvature c. The hypersurface M is said to satisfy *geodesic chord property* if the geodesic chord in the ambient space $\bar{M}^n(c)$ joining any two points on M meets the hypersurface at the same angle at the two points. Note that a geodesic chord in the ambient space $\bar{M}^n(c)$ is defined by a segment of a geodesic of $\bar{M}^n(c)$ with two end points. When $c > 0$, that is, the ambient space $\bar{M}^n(c)$ is a hypersphere S^n in the Euclidean $(n+1)$-space $\mathbb{E}^{n+1}$, we consider only two points which are not antipodal to each other.

First of all, in Section 2, we study spherical hypersurfaces in the n-dimensional unit sphere S^n which satisfy the geodesic chord property and then classify such hypersurfaces. Next, in Section 3, we study and classify completely the hypersurfaces in the n-dimensional hyperbolic space H^n with the geodesic chord property, which is imbedded in the Minkowski space $\mathbb{E}_1^{n+1}$.

In the Euclidean space $\mathbb{E}^{n+1}$, some characterizations of hyperspheres, ellipsoids, elliptic paraboloids and elliptic hyperboloids were given in [7–10], respectively. In the Minkowski space $\mathbb{E}_1^{n+1}$, a few characterizations of hyperbolic spaces were also established in [11].

Throughout this article, we assume that all objects are smooth and connected unless otherwise mentioned.

2. Spherical Hypersurfaces

In this section, we consider a hypersurface M in the unit hypersphere $S^n \subset \mathbb{E}^{n+1}$ centered at the origin.

For each point $x \in M$, we denote by $G(x)$ the unit normal to $T_x M$ in the unit hypersphere S^n. Then $G : M \to S^n$ is called the Gauss map of M in the unit hypersphere S^n. For an orthonormal local frame field $\{e_1(x), \ldots, e_{n-1}(x)\}$ on M, $\{e_1(x), \ldots, e_{n-1}(x), G(x), x\}$ forms an orthonormal basis of $T_x \mathbb{E}^{n+1}$. We denote by $A(x)$ the $(n+1) \times (n-1)$ frame matrix $[e_1(x), \ldots, e_{n-1}(x)]$ with column vectors $e_1(x), \ldots, e_{n-1}(x)$.

For any two points $x, y \in M \subset S^n$ with $x + y \neq 0$, we denote by $\theta \in (0, \pi)$ the angle between x and y in the ambient space $\mathbb{E}^{n+1}$. Then the unit speed geodesic chord $\gamma(t)$ with $\gamma(0) = x$ and $\gamma(\theta) = y$ is given by $\gamma(t) = \cos tx + \sin tv$, where we put

$$v = v_{x,y} = -\cot\theta x + \csc\theta y. \tag{1}$$

The geodesic chord has initial velocity $v = v_{x,y}$ at x.

The projection $P_x(v)$ of v onto the tangent space $T_x M$ is given by

$$P_x(v) = A(x)\alpha, \tag{2}$$

where α is a vector in $\mathbb{E}^{n-1}$. In order to determine $\alpha \in \mathbb{E}^{n-1}$, first note that $v - P_x(v)$ is perpendicular to $T_x M$. Then, (2) shows that

$$A(x)^t(v - A(x)\alpha) = 0, \tag{3}$$

where $A(x)^t$ denotes the transpose of $A(x)$. Since $A(x)^t A(x)$ is the $(n-1) \times (n-1)$ identity matrix I_{n-1}, we get from (3)

$$\alpha = A(x)^t v. \tag{4}$$

Thus, the projection $P_x(v)$ of v into the tangent space $T_x M$ is given by

$$P_x(v) = P(x)v, \tag{5}$$

where $P(x)$ denotes the $(n+1) \times (n+1)$ matrix given by

$$P(x) = A(x)A(x)^t. \tag{6}$$

Now, for later use we prove a lemma as follows.

Lemma 1. *For a spherical hypersurface M, the following are equivalent:*

1. *M satisfies the geodesic chord property.*
2. *For points $x, y \in M$, we have*

$$|A(x)^t y| = |A(y)^t x|. \tag{7}$$

3. *For points $x, y \in M$, we have*

$$\langle G(x), y \rangle = \epsilon \langle x, G(y) \rangle, \tag{8}$$

where $G(x)$ denotes the Gauss map of M and $\epsilon = \pm 1$.

Proof. For two points $x, y \in M \subset S^n$ with $x + y \neq 0$, we denote by θ the angle between x and y as above. If we let ϕ denote the angle between the geodesic chord γ from x to y and the tangent plane $T_x M$ at the point x, then together with (1), (5) and (6) show that

$$\cos \phi = |P(x)v| = \csc \theta |P(x)y|, \tag{9}$$

where the second equality follows from $A(x)^t x = 0$, because x is orthogonal to the tangent plane $T_x M$. Using $A(x)^t A(x) = I_{n-1}$, we see that

$$|P(x)y| = |A(x)^t y|. \tag{10}$$

Hence we have

$$\cos \phi = \csc \theta |A(x)^t y|. \tag{11}$$

Similarly, by interchanging x and y in the above discussions the angle ψ between the geodesic chord from y to x and the tangent plane $T_y M$ at the point y is given by

$$\cos \psi = |P(y)v_{y,x}| = \csc \theta |A(y)^t x|, \tag{12}$$

where we use

$$v_{y,x} = -y \cot \theta + x \csc \theta. \tag{13}$$

Together with (10)–(12) imply that (1) and (2) in Lemma 1 are equivalent to each other for $x, y \in M$ with $x + y \neq 0$. By continuity, (8) holds for all $x, y \in M$.

If we consider the following expression of y

$$y = \sum_{i=1}^{n-1} y_i e_i(x) + y_n G(x) + y_{n+1} x, \tag{14}$$

then we have

$$y_i = \langle y, e_i(x) \rangle, \quad i = 1, 2, \ldots, n-1 \tag{15}$$

and

$$y_n = \langle y, G(x) \rangle, \quad y_{n+1} = \langle y, x \rangle. \tag{16}$$

Hence we get from (14)–(16)

$$|A(x)^t y|^2 = \sum_{i=1}^{n-1} y_i^2 = |y|^2 - \langle y, G(x) \rangle^2 - \langle y, x \rangle^2, \tag{17}$$

where the first equality follows from $A(x)^t y = (y_1, \ldots, y_{n-1})^t$.

By interchanging x and y, we obtain from (17)

$$|A(y)^t x|^2 = |x|^2 - \langle x, G(y) \rangle^2 - \langle x, y \rangle^2. \tag{18}$$

Combining (17) and (18) shows that (2) and (3) are equivalent to each other. This completes the proof of Lemma 1. $\square$

Remark 1. *Without using (7), we may prove the equivalence of (1) and (3) in Lemma 1. Since (7) holds for spherical submanifolds with geodesic chord property, it is useful in the study of such spherical submanifolds.*

We now suppose that a hypersurface M in the unit hypersphere $S^n \subset \mathbb{E}^{n+1}$ satisfies the geodesic chord property. We may assume, without loss of generality, that M lies fully in the Euclidean space $\mathbb{E}^{n+1}$, which means that M is not contained in any hyperplane of $\mathbb{E}^{n+1}$. Otherwise, it is an open part of a small sphere $S^{n-1}(r) \subset S^n$ for some $r \in (0, 1]$. Hence, on M there exist points $y_0, y_1, \ldots, y_{n+1}$ such that the set $\{A_j | A_j = y_j - y_0, j = 1, 2, \ldots, n+1\}$ spans the Euclidean space $\mathbb{E}^{n+1}$.

It follows from (3) of Lemma 1 that we have

$$\langle G(x), y_0 \rangle = \epsilon \langle G(y_0), x \rangle \tag{19}$$

and

$$\langle G(x), y_j \rangle = \epsilon \langle G(y_j), x \rangle, j = 1, 2, \ldots, n+1, \tag{20}$$

where $\epsilon = \pm 1$. Hence we obtain from (19) and (20)

$$\langle G(x), A_j \rangle = \langle B_j, x \rangle, j = 1, 2, \ldots, n+1, \tag{21}$$

where we put for $j = 1, 2, \ldots, n+1$

$$A_j = y_j - y_0, B_j = \epsilon G(y_j) - \epsilon G(y_0). \tag{22}$$

We denote by A the $(n+1) \times (n+1)$ matrix defined by

$$A^t = [B_1, B_2, \ldots, B_{n+1}][A_1, A_2, \ldots, A_{n+1}]^{-1}, \tag{23}$$

where $[A_1, A_2, \ldots, A_{n+1}]$ is the $(n+1) \times (n+1)$ matrix consisting of columns $A_1, A_2, \ldots, A_{n+1}$, etc. Then we have from (21)

$$G(x) = Ax. \tag{24}$$

Therefore, we get the following lemma.

Lemma 2. *Suppose that a hypersurface M in the unit hypersphere $S^n \subset \mathbb{E}^{n+1}$ satisfies the geodesic chord property. If the hypersurface M lies fully in $\mathbb{E}^{n+1}$, then for an $(n+1) \times (n+1)$ matrix A, the Gauss map $G(x)$ satisfies $G(x) = Ax$.*

Remark 2. *If the hypersurface M does not lie fully in $\mathbb{E}^{n+1}$, that is, M is contained in a hyperplane, then it is an open part of a small sphere $S^{n-1}(r) \subset S^n$ for some $r \in (0, 1]$. In fact, for some unit vector $a \in \mathbb{E}^{n+1}$ we have*

$$M \subset \{x \in S^n | \langle x, a \rangle = \cos \theta\}, \tag{25}$$

where $\cos \theta = \sqrt{1 - r^2}$. Then, we get

$$G(x) = -\cot \theta x + \csc \theta a. \tag{26}$$

Thus, M satisfies $G(x) = Ax + b$ for $A = -\cot \theta I$ and $b = \csc \theta a$.

Finally, we need the following proposition, which was proved in [12].

Proposition 2. *A hypersurface M in the unit hypersphere $S^n \subset \mathbb{E}^{n+1}$ satisfies for an $(n+1) \times (n+1)$ matrix A and a vector $b \in \mathbb{E}^{n+1}$*

$$G(x) = Ax + b \tag{27}$$

if and only if it is an open part of either a sphere $S^{n-1}(r)$ or a product $S^p(r_1) \times S^{n-p-1}(r_2)$ of spheres with $r_1^2 + r_2^2 = 1$.

It follows from Proposition 2 that the hypersurface M is an open part of either a sphere $S^{n-1}(r)$ or a product $S^p(r_1) \times S^{n-p-1}(r_2)$ of spheres with $r_1^2 + r_2^2 = 1$.

Conversely, if M is a small sphere $S^{n-1}(r) \subset S^n$ for some $r \in (0,1]$, then together with (26), Lemma 1 shows that M satisfies the geodesic chord property. If M is the product $S^p(r_1) \times S^{n-p-1}(r_2) \subset \mathbb{E}^{p+1} \times \mathbb{E}^{n-p}$ with $r_1^2 + r_2^2 = 1$, then at $x = (x_1, x_2) \in \mathbb{E}^{p+1} \times \mathbb{E}^{n-p}$ we have

$$G(x_1, x_2) = \frac{1}{\sqrt{r_1^2 + r_2^2}} \left(\frac{r_2}{r_1} x_1, -\frac{r_1}{r_2} x_2 \right). \tag{28}$$

Hence, it follows from Lemma 1 that M satisfies the geodesic chord property.

Summarizing the above discussions, we get the classification theorem as follows.

Theorem 1. *For a hypersurface M in the unit hypersphere $S^n \subset \mathbb{E}^{n+1}$, the following are equivalent:*

1. *M satisfies the geodesic chord property.*
2. *The Gauss map G satisfies $|\langle G(x), y \rangle| = |\langle G(y), x \rangle|$ for arbitrary $x, y \in M$.*
3. *The Gauss map G satisfies $G(x) = Ax + b$ for an $(n+1) \times (n+1)$ matrix A and a vector $b \in \mathbb{E}^{n+1}$.*
4. *M is an open portion of either a sphere $S^{n-1}(r)$ or a product $S^p(r_1) \times S^{n-p-1}(r_2)$ of spheres with $r_1^2 + r_2^2 = 1$.*

3. Hypersurfaces in the Hyperbolic Space

In this section, we consider a hypersurface M in the hyperbolic space H^n which lies in the $(n+1)$-dimensional Minkowski space $\mathbb{E}_1^{n+1}$.

First of all, let us recall some preliminaries. We consider the $(n+1)$-dimensional Minkowski space $\mathbb{E}_1^{n+1}$ with metric $ds^2 = dx_1^2 + \cdots + dx_n^2 - dx_{n+1}^2$, where $x = (x_1, \cdots, x_{n+1})$. In other words, for $x, y \in \mathbb{E}_1^{n+1}$ we use the Lorentzian scalar product $\langle x, y \rangle_1 = x_1 y_1 + \cdots + x_n y_n - x_{n+1} y_{n+1}$. Let us denote by $H^n(r) \subset \mathbb{E}_1^{n+1}$ the spacelike hyperquadric defined by $\langle x, x \rangle_1 = -r^2$ with $x_{n+1} > 0$. Then $H^n(r)$ is a Riemannian space form with constant sectional curvature $K = -\frac{1}{r^2}$. When $r = 1$, $H^n(1)$ is called the standard imbedding of the hyperbolic space H^n of curvature $K = -1$, or simply the hyperbolic space (cf. [13,14]).

We introduce a notation for the Lorentzian scalar product. For a vector $v = (v_1, \ldots, v_{n+1}) \in \mathbb{E}_1^{n+1}$, we put $\bar{v} = (v_1, \ldots, v_n, -v_{n+1})$. Then for any $v, w \in \mathbb{E}_1^{n+1}$, we have $\langle v, w \rangle_1 = \langle v, \bar{w} \rangle$, where $\langle \cdot, \cdot \rangle$ denotes the Euclidean scalar product. For an $(n+1) \times k$ matrix $A = [A_1, \ldots, A_k]$ with column vectors $A_1, \ldots, A_k$, we let $\bar{A} = [\bar{A}_1, \ldots, \bar{A}_k]$. Then we have $(\bar{A}v) = \bar{A}v$ for any k-dimensional vector v.

For a point $x \in M$, we denote by $G(x)$ the unit normal to $T_x M$ in the hyperbolic space H^n. Then G is called the Gauss map of M in the hyperbolic space H^n. For an orthonormal local frame field $\{e_1(x), \ldots, e_{n-1}(x)\}$ on M, $\{e_1(x), \ldots, e_{n-1}(x), G(x), x\}$ forms an orthonormal basis of $T_x \mathbb{E}_1^{n+1}$ with respect to the Lorentzian scalar product. We denote by $A(x)$ the $(n+1) \times (n-1)$ frame matrix $[e_1(x), \ldots, e_{n-1}(x)]$ with column vectors $e_1(x), \ldots, e_{n-1}(x)$. Then, we have for the frame matrix $A(x)$

$$A(x)^t \bar{A}(x) = I_{n-1} \tag{29}$$

and

$$A(x)\bar{x} = \bar{A}(x)x = 0. \tag{30}$$

For any two points $x, y \in M \subset H^n$, we have $\langle x, y \rangle_1 = \langle x, \bar{y} \rangle < -1$. Hence $\langle x, y \rangle_1 = -\cosh\theta$ for some positive θ, which is called the *hyperbolic angle* between x and y ([14], p. 144). Then the unit speed geodesic chord $\gamma(t)$ with $\gamma(0) = x$ and $\gamma(\theta) = y$ is given by $\gamma(t) = \cosh tx + \sinh tv$, where

$$v = v_{x,y} = -\coth\theta x + \operatorname{csch}\theta y. \tag{31}$$

Note that the geodesic chord $\gamma(t)$ has initial velocity $v = v_{x,y}$ at x. The angle between v and the tangent space $T_x M \subset T_x H^n$ is defined by the angle between v and the projection $P_x(v)$ of v into $T_x M$. Hence, if we let ϕ denote the angle between the geodesic chord γ from x to y and the tangent plane $T_x M$ at the point x, then we have

$$\cos \phi = |P_x(v)| = \sqrt{\langle P_x(v), P_x(v) \rangle_1}. \tag{32}$$

By a similar argument to that in Section 2, using (29)–(32) we may prove the following two lemmas. We omit the proofs.

Lemma 3. *For a hypersurface M in the hyperbolic space H^n, the following are equivalent:*

1. *M satisfies the geodesic chord property.*
2. *For any two points $x, y \in M$, the frame matrix A of M satisfies*

$$\langle A(x)^t \bar{y}, A(x)^t \bar{y} \rangle = \langle A(y)^t \bar{x}, A(y)^t \bar{x} \rangle, \tag{33}$$

   where $\langle \cdot, \cdot \rangle$ is the Euclidean inner product.
3. *For any two points $x, y \in M$, the Gauss map $G(x)$ of M satisfies*

$$\langle G(x), y \rangle_1 = \epsilon \langle x, G(y) \rangle_1, \tag{34}$$

   where $\epsilon = \pm 1$.

Lemma 4. *Suppose that a hypersurface M in the hyperbolic space $H^n \subset \mathbb{E}_1^{n+1}$ satisfies the geodesic chord property. If the hypersurface M is full in $\mathbb{E}_1^{n+1}$, then for an $(n+1) \times (n+1)$ matrix A, the Gauss map $G(x)$ satisfies $G(x) = Ax$.*

If a hypersurface $M \subset H^n$ is not full in $\mathbb{E}_1^{n+1}$, then M is contained in a hyperplane $P = \{x \in \mathbb{E}_1^{n+1} | \langle x, a \rangle = c\}$ for some nonzero $a \in \mathbb{E}_1^{n+1}$ and some constant c.

We divide by three cases according to the causal character of the nonzero vector $a \in \mathbb{E}_1^{n+1}$.

Case 1. Suppose that $\langle a, a \rangle < 0$. Then, up to congruences of H^n we may assume that $a = (0, \ldots, 0, 1)$. Hence we have for some θ

$$M \subset \{x \in H^n | x_{n+1} = \cosh \theta\}. \tag{35}$$

Thus, M is an open part of the hypersphere $S^{n-1}(\sinh \theta)$ in the Euclidean space $\mathbb{E}^n \subset \mathbb{E}_1^{n+1}$. In this case, we have

$$G(x) = -\coth \theta x + \operatorname{csch}\theta a. \tag{36}$$

This shows that M satisfies $G(x) = Ax + b$ for $A = -\coth \theta I$ and $b = \operatorname{csch}\theta a$.

Case 2. Suppose that $\langle a, a \rangle > 0$. Then, up to congruences of H^n we may assume that $a = (1, 0, \ldots, 0)$. Hence we have for some θ

$$M \subset \{x \in H^n | x_1 = \sinh \theta\}. \tag{37}$$

Thus, M is an open part of the hyperbolic space $H^{n-1}(\cosh \theta)$ in the Minkowski space $\mathbb{E}_1^n \subset \mathbb{E}_1^{n+1}$. In this case, we have

$$G(x) = \tanh \theta x + \operatorname{sech}\theta a, \tag{38}$$

which shows that M satisfies $G(x) = Ax + b$ for $A = \tanh \theta I$ and $b = \operatorname{sech}\theta a$.

Case 3. Suppose that $\langle a, a \rangle = 0$. Then, up to congruences of H^n we may assume that $a = (0, \ldots, 0, 1, 1)$ and $c = -1$. Hence we have

$$M \subset N = \{x \in H^n | x_{n+1} = x_n + 1\}. \tag{39}$$

Note that $N = \{(x, f(x), f(x) + 1) | x \in \mathbb{E}^{n-1}\} \subset H^n$, where $f(x) = \frac{1}{2}|x|^2$ for $x = (x_1, \ldots, x_{n-1}) \in \mathbb{E}^{n-1}$. In this case, we have

$$G(x) = x - a,\tag{40}$$

which shows that M satisfies $G(x) = Ax + b$ for $A = I$ and $b = -a$.

Finally, we use the following proposition ([12]).

Proposition 3. *A hypersurface M in $H^n \subset \mathbb{E}_1^{n+1}$ satisfies $G(x) = Ax + b$ if and only if M is isoparametric, or equivalently M is an open piece of one of the following hypersurfaces:*

1. $S^{n-1}(\sinh\theta) \subset H^n$,
2. $H^{n-1}(\cosh\theta) \subset H^n$,
3. $S^p(\sinh\theta) \times H^{n-p-1}(\cosh\theta) \subset H^n$,
4. $N = \{(x, f(x), f(x) + 1) | f(x) = \frac{1}{2}|x|^2, x \in \mathbb{E}^{n-1}\} \subset H^n$.

Summarizing the above discussions, we prove the following classification theorem.

Theorem 2. *A hypersurface M in $H^n \subset \mathbb{E}_1^{n+1}$ satisfies the geodesic chord property if and only if it is an open piece of one of the hypersurfaces in Proposition 3.*

Proof. Together with Lemma 3, it follows from (36), (38) and (39) that the hypersurfaces $S^{n-1}(\sinh\theta)$, $H^{n-1}(\cosh\theta)$ and N satisfies the geodesic chord property, respectively. Hence, it remains to show that $S^p(\sinh\theta) \times H^{n-p-1}(\cosh\theta) \subset H^n$ satisfies the geodesic chord property.

Suppose that $M = S^p(\sinh\theta) \times H^{n-p-1}(\cosh\theta)$. Then, for a point $x = (x_1, x_2) \in S^p(\sinh\theta) \times H^{n-p-1}(\cosh\theta)$ we have

$$G(x) = (\coth\theta x_1, \tanh\theta x_2).\tag{41}$$

Thus, Lemma 3 shows that M satisfies the geodesic chord property. This completes the proof. □

From the proof of Theorem 2, we also obtain

Theorem 3. *For a hypersurface M in the hyperbolic space $H^n \subset \mathbb{E}_1^{n+1}$, the following are equivalent:*

1. *M satisfies the geodesic chord property.*
2. *The Gauss map G satisfies $|\langle G(x), y\rangle_1| = |\langle G(y), x\rangle_1|$ for any $x, y \in M$.*
3. *The Gauss map G satisfies $G(x) = Ax + b$ for an $(n+1) \times (n+1)$ matrix A and a vector $b \in \mathbb{E}_1^{n+1}$.*
4. *M is an isoparametric hypersurface of H^n.*
5. *M is an open part of one of the following hypersurfaces: $S^{n-1}(r)$, $H^{n-1}(r)$, $S^p(r_1) \times H^{n-p-1}(r_2)$, N, where $r_2^2 - r_1^2 = 1$ and $N = \{(x, \frac{1}{2}|x|^2, \frac{1}{2}|x|^2 + 1) | x \in \mathbb{E}^{n-1}\}$.*

4. Conclusions

In this paper, we have classified hypersurfaces in the nonflat space forms satisfying the so-called geodesic chord property. As a result, we have shown that such a spherical hypersurface is an open portion of either a sphere or a product of two spheres, which are isoparametric spherical ones with (at most) two principal curvatures. For hypersurfaces in the hyperbolic space, we have proven that the geodesic chord property is another characterization of isoparametric ones. We hope the results will help studying hypersurfaces in the nonflat space forms.

Author Contributions: D.-S.K. and Y.H.K. set up the problem and computed the details and D.W.Y. checked and polished the draft. These authors contributed equally to this work.

Funding: This research was supported by Basic Science Research Program through the National Research Foundation of Korea (NRF) funded by the Ministry of Education (NRF-2018R1D1A3B05050223).

Acknowledgments: We are very thankful to the reviewers for their suggestions to improve the quality of this paper.

Conflicts of Interest: The authors declare no conflict of interest.

References

1. Rademacher, H.; Toeplitz, O. *The Enjoyment of Mathematics*; Translated from the Second (1933) German Edition and with Additional Chapters by H. Zuckerman; Princeton Science Library, Princeton University Press: Princeton, NJ, USA, 1994.
2. Chen, B.-Y.; Kim, D.-S.; Kim, Y.H. New characterization of W-curves. *Publ. Math. Debr.* **2006**, *69*, 457–472.
3. Kim, D.-S.; Kim, Y.H. New characterizations of spheres, cylinders and W-curves. *Linear Algebra Appl.* **2010**, *432*, 3002–3006. [CrossRef]
4. Boas, H.P. A geometric characterization of the ball and the Bochner-Martinelli kernel. *Math. Ann.* **1980**, *248*, 275–278. [CrossRef]
5. Boas, H.P. Spheres and cylinders: A local geometric characterization. *Ill. J. Math.* **1984**, *28*, 120–124. [CrossRef]
6. Wegner, B. A differential geometric proof of the local geometric characterization of spheres and cylinders by Boas. *Math. Balk. (N.S.)* **1988**, *2*, 294–295.
7. Kim, D.-S. Ellipsoids and Elliptic hyperboloids in the Euclidean space E^{n+1}. *Linear Algebra Appl.* **2015**, *471*, 28–45. [CrossRef]
8. Kim, D.-S.; Kim, Y.H. Some characterizations of spheres and elliptic paraboloids. *Linear Algebra Appl.* **2012**, *437*, 113–120. [CrossRef]
9. Kim, D.-S.; Kim, Y.H. Some characterizations of spheres and elliptic paraboloids II. *Linear Algebra Appl.* **2013**, *438*, 1356–1364. [CrossRef]
10. Kim, D.-S.; Song, B. A characterization of elliptic hyperboloids. *Honam Math. J.* **2013**, *35*, 37–49. [CrossRef]
11. Kim, D.-S.; Kim, Y.H.; Yoon, D.W. On standard imbeddings of hyperbolic spaces in the Minkowski space. *Comptes Rendus Math.* **2014**, *352*, 1033–1038. [CrossRef]
12. Kim, D.-S. On the Gauss map of Hypersurfaces in the space form. *J. Korean Math. Soc.* **1995**, *32*, 509–518.
13. Kühnel, W. *Differential Geometry, Curves-Surfaces-Manifolds*; Translated from the 1999 German Original by Bruce Hunt; Student Mathematical Library, 16; American Mathematical Society: Providence, RI, USA, 2002.
14. O'Neill, B. *Semi-Riemannian Geometry with Applications to Relativity*; Pure and Applied Mathematics, 103; Academic Press, Inc.: New York, NY, USA, 1983.

symmetry

MDPI

Article

New Characterizations of the Clifford Torus and the Great Sphere

Sun Mi Jung [1], Young Ho Kim [1,*] and Jinhua Qian [2]

[1] Department of Mathematics, Kyungpook National University, Daegu 41566, Korea
[2] Department of Mathematics, Northeastern University, Shenyang 110004, China
* Correspondence: yhkim@knu.ac.kr; Tel.: +82-53-950-7325

Received: 2 August 2019; Accepted: 23 August 2019; Published: 27 August 2019

Abstract: In studying spherical submanifolds as submanifolds of a round sphere, it is more relevant to consider the spherical Gauss map rather than the Gauss map of those defined by the oriented Grassmannian manifold induced from their ambient Euclidean space. In that sense, we study ruled surfaces in a three-dimensional sphere with finite-type and pointwise 1-type spherical Gauss map. Concerning integrability and geometry, we set up new characterizations of the Clifford torus and the great sphere of 3-sphere and construct new examples of spherical ruled surfaces in a three-dimensional sphere.

Keywords: Clifford torus; spherical Gauss map; finite-type; pointwise 1-type spherical Gauss map

1. Introduction

In the 1960s, T. Takahashi proved that an isometric immersion $x : M \to \mathbb{E}^m$ of a Riemannian manifold M into a Euclidean space $\mathbb{E}^m$ satisfies $\Delta x = \lambda x$ ($\lambda \neq 0$) if, and only if, it is part of a hypersphere or a minimal submanifold of a hypersphere, where Δ denotes the Laplacian of M [1]. Generalizing such an eigenvalue problem of immersion, B.-Y. Chen introduced the notion of finite-type immersion of a Riemannian manifold M into a Euclidean space $\mathbb{E}^m$ in the late 1970s. Since then, it has been used as a remarkably useful tool in differential geometry to classify and characterize many manifolds including minimal submanifolds in $\mathbb{E}^m$. In particular, minimal submanifolds of Euclidean space are considered as a spacial case of submanifolds of the finite-type, in fact they are of 1-type [2,3]. Thanks to Nash's embedding theorem of Riemannian manifolds, it has been a natural consideration of Riemannian manifolds as submanifolds in Euclidean space along with the notion of finite-type immersion.

A ruled surface or a ruled submanifold of Euclidean space or Minkowski space is one of the most natural geometric objects in classical differential geometry which has been examined under finite-type related geometric conditions [4–7]. The well-known Catalan's Theorem says that the only minimal ruled surfaces in Euclidean 3-space are the planes and the helicoids. A general ruled submanifold of a smooth manifold is defined by a foliation of totally geodesic submanifolds along a smooth curve. In [8], it was shown that a regular and connected ruled surface M in $\mathbb{S}^3$ is of finite-type if and only if it is an open part of a ruled minimal surface in $\mathbb{S}^3$ or an open part of a Riemannian product of two circles of different radii.

Such a theory of finite-type immersion in a Riemannian sense was naturally extended to an isometric immersion of a manifold M into a pseudo-Euclidean space $\mathbb{E}^m_s$ with index s and the smooth functions defined on a submanifold in $\mathbb{E}^m$ or $\mathbb{E}^m_s$. In particular, the Gauss map on a submanifold in $\mathbb{E}^m$ or $\mathbb{E}^m_s$ is the most interesting and useful object which involves rich geometrical and topological properties on the submanifold.

Regarding the Gauss map of finite-type, B.-Y. Chen and P. Piccinni initiated the study of submanifolds with a finite-type Gauss map in Euclidean space [9]. Many works about submanifolds in $\mathbb{E}^m$ or $\mathbb{L}^m$ with a finite-type Gauss map have been achieved [9–13]. In [10], C. Baikoussis showed that the only ruled submanifolds M^{n+1} in Euclidean space $\mathbb{E}^m$ with a finite-type Gauss map are the cylinders over curves of finite-type and the $(n+1)$-dimensional Euclidean spaces. Ruled surfaces and ruled submanifolds with a finite-type Gauss map in Minkowski space were examined and completely classified in [6,14–17].

During the last ten years or so, the present authors et. al have worked on submanifolds of Euclidean or pseudo-Euclidean space which look similar to those of 1-type Gauss maps, which is called pointwise 1-type. For example, the Gauss maps G of the helicoid and the right cone in $\mathbb{E}^3$ satisfy $\Delta G = f(G + \mathbf{C})$ for a nonzero smooth function f and a constant vector $\mathbf{C}$ (cf. [18–20]) . Since it was introduced in [18], many works concerning pointwise 1-type Gauss maps were established in [19–24]. In [22], the authors showed that the ruled submanifold M in $\mathbb{E}^m$ is minimal if, and only if, the Gauss map G of M is pointwise 1-type of the first kind. The classification theorems of ruled submanifolds in the Euclidean space $\mathbb{E}^m$ and the Minkowski space $\mathbb{L}^m$ with pointwise 1-type Gauss maps were completed [25,26].

On the other hand, one of the important manifolds in differential geometry is a sphere or a spherical submanifold. Regarding such manifolds, Obata studied the spherical Gauss map for a spherical submanifold M in the unit hypersphere $\mathbb{S}^m$ ($\subset \mathbb{E}^{m+1}$) [27]. The set S of all the great n-spheres in $\mathbb{S}^m$ is naturally identified with the oriented Grassmannian manifold of $(n+1)$-planes through the center of $\mathbb{S}^m$ in $\mathbb{E}^{m+1}$ because such $(n+1)$-planes determine unique great n-spheres and conversely [27]: A spherical Gauss map of an immersion x of a Riemannian manifold M into $\mathbb{S}^m$ is a map of M into the oriented Grassmannian manifold $G(n+1, m+1)$ which assigns to each point p of M the great n-sphere tangent to M at $x(p)$, or the $(n+1)$-plane spanned by the tangent space of M at $x(p)$ and the normal to $\mathbb{S}^m$ at $x(p)$ in $\mathbb{E}^{m+1}$. Granted, the spherical Gauss map is more meaningful than the classical Gauss map in the study of spherical submanifolds (cf. [28,29]). Extending the notion of finite-type Gauss maps of submanifolds of Euclidean space in the usual sense, B.-Y. Chen and H.-S. Lue initiated the study of spherical submanifolds with finite-type spherical Gauss maps and obtained several fundamental results in this respect [28]. Recently, some works on spherical submanifolds with low-type spherical or pseudo-spherical Gauss maps have been made [30–32].

In this article, we study ruled surfaces in $\mathbb{S}^3$ by means of the spherical Gauss map to characterize the Clifford torus and the great sphere in the three-dimensional unit sphere $\mathbb{S}^3$.

In the present paper, all geometric objects are assumed to be smooth, and manifolds under consideration are connected unless otherwise stated.

2. Preliminaries

Let $x : M \to \mathbb{S}^{m-1}$ be an isometric immersion of an n-dimensional Riemannian manifold M into a unit sphere $\mathbb{S}^{m-1}(\subset \mathbb{E}^m)$. We identify x with its position in a vector field. Let $(y_1, y_2, \ldots, y_m)$ be a local coordinate system of M in $\mathbb{S}^{m-1}$. For the components g_{ij} of the Riemannian metric $\langle \cdot, \cdot \rangle$ on M induced from that of $\mathbb{S}^{m-1}$, we denote by (g^{ij}) (respectively, $\mathcal{G}$) the inverse matrix (respectively, the determinant) of the matrix (g_{ij}). Then the Laplace operator Δ on M is defined by

$$\Delta = -\frac{1}{\sqrt{\mathcal{G}}} \sum_{i,j} \frac{\partial}{\partial y_i} \left(\sqrt{\mathcal{G}} g^{ij} \frac{\partial}{\partial y_j} \right).$$

An immersion x of a manifold M into $\mathbb{S}^{m-1}$ is said to be of finite-type if its position vector field x can be expressed as a finite sum of spectral decomposition as follows

$$x = x_0 + x_1 + \cdots + x_k$$

for some positive integer k, where x_0 is a constant vector, and $\Delta x_i = \lambda_i x_i$ for some $\lambda_i \in \mathbb{R}$, $i = 1, \ldots, k$. If $\lambda_1, \ldots, \lambda_k$ are mutually different, M is said to be of k-type. Similarly, a smooth map ϕ on an n-dimensional submanifold M of $\mathbb{S}^{m-1}$ is said to be of *finite-type* if ϕ is a finite sum of $\mathbb{E}^m$-valued eigenfunctions of Δ. In particular, we say that a smooth map ϕ is *harmonic* if $\Delta\phi = 0$. If the manifold M is compact without boundary, a harmonic map is constant and thus it is of finite-type. In general, harmonic smooth map is not necessarily of finite-type if M is not compact.

Let Π be an oriented n-plane in $\mathbb{E}^m$ and $e_1, ..., e_n$ an orthonormal basis of Π. If we identify an oriented n-plane Π with a decomposable n vector $e_1 \wedge \cdots \wedge e_n$ defined by the exterior algebra in a natural way, the oriented Grassmannian manifold $G(n, m)$ can be regarded as the set of all oriented n-planes in $\mathbb{E}^N = \Lambda^n \mathbb{E}^m$, where $N = \binom{m}{n}$. Moreover, we can define an inner product in $G(n, m)$ by

$$\ll e_{i_1} \wedge \cdots \wedge e_{i_n}, e_{j_1} \wedge \cdots \wedge e_{j_n} \gg = \det(\langle e_{i_l}, e_{j_k} \rangle)$$

for two vectors $e_{i_1} \wedge \cdots \wedge e_{i_n}$ and $e_{j_1} \wedge \cdots \wedge e_{j_n}$ in $\mathbb{E}^N$.

From now on we assume that the unit sphere $\mathbb{S}^{m-1}$ is centered at the origin in $\mathbb{E}^m$. We identify each tangent vector X of M in $\mathbb{S}^{m-1}$ with the differential $dx(X)$.

For a spherical submanifold M in $\mathbb{S}^{m-1}$, the position vector x of each point p of $\mathbb{S}^{m-1}$ and an orthonormal basis $\{e_1, e_2, ..., e_n\}$ of the tangent space T_pM determine an oriented $(n+1)$-plane in $\mathbb{E}^m$. Thus, we can have a map

$$G : M \to G(n+1, m)$$

via $G(p) = x \wedge e_1 \wedge \cdots \wedge e_n$. We call G the *spherical Gauss map* of M in $\mathbb{S}^{m-1}$. This map can be viewed as

$$G : M \to G(n+1, m) \subset \mathbb{S}^{\binom{m}{n+1}-1} \subset \mathbb{E}^{\binom{m}{n+1}}$$

by considering the norm of vectors. We now define the pointwise 1-type spherical Gauss map of the spherical submanifold.

Definition 1. *An oriented n-dimensional submanifold M of $\mathbb{S}^{m-1}$ is said to have pointwise 1-type spherical Gauss map G if it satisfies the partial differential equation*

$$\Delta G = f(G + C) \tag{1}$$

for a nonzero smooth function f on M and some constant vector C. In particular, if C is zero, the spherical Gauss map G is said to be pointwise 1-type of the first kind. Otherwise, it is said to be of the second kind.

3. Ruled Surfaces in $\mathbb{S}^3$ with Harmonic Spherical Gauss Maps

Let M be a ruled surface in the sphere $\mathbb{S}^3$ ($\subset \mathbb{E}^4$). Then, it is foliated by geodesics of $\mathbb{S}^3$ along a spherical curve. So, we can put its parametrization with spherical curves $\alpha = \alpha(s)$ and $\beta = \beta(s)$ by

$$x = x(s, t) = \cos t\, \alpha(s) + \sin t\, \beta(s), \quad s \in I, \ t \in J, \tag{2}$$

where I and J are some open intervals. Without loss of generality, we may assume that

$$\langle \alpha, \alpha \rangle = \langle \beta, \beta \rangle = \langle \alpha', \alpha' \rangle = 1 \quad \text{and} \quad \langle \alpha, \beta \rangle = \langle \alpha', \beta \rangle = 0. \tag{3}$$

From now on, we always assume that the Parametrization (2) satisfies Condition (3) unless otherwise stated. Then, the spherical Gauss map G of M is given by

$$G = \frac{x \wedge x_s \wedge x_t}{||x \wedge x_s \wedge x_t||} = \frac{1}{\sqrt{q}}\Big(\cos t\, \alpha(s) \wedge \alpha'(s) \wedge \beta(s) + \sin t\, \alpha(s) \wedge \beta'(s) \wedge \beta(s)\Big), \tag{4}$$

where the function $q = q(s, t)$ is defined by

$$q = \langle x_s, x_s \rangle = \cos^2 t + 2u(s) \cos t \sin t + w(s) \sin^2 t, \tag{5}$$

where $u(s) = \langle \alpha', \beta' \rangle$ and $w(s) = \langle \beta', \beta' \rangle$ are functions of s.

By the definition of the Laplace operator Δ, we have

$$
\begin{aligned}
\Delta G = &- q^{-\frac{7}{2}}(q_s)^2(\cos tA + \sin tB) + \frac{3}{2}q^{-\frac{5}{2}}q_s(\cos tA' + \sin tB') \\
&+ \frac{1}{2}q^{-\frac{5}{2}}q_{ss}(\cos tA + \sin tB) - q^{-\frac{3}{2}}(\cos tA'' + \sin tB'') \\
&- \frac{1}{2}q^{-\frac{5}{2}}(q_t)^2(\cos tA + \sin tB) + \frac{1}{2}q^{-\frac{3}{2}}q_t(-\sin tA + \cos tB) \\
&+ \frac{1}{2}q^{-\frac{3}{2}}q_{tt}(\cos tA + \sin tB) + q^{-\frac{1}{2}}(\cos tA + \sin tB),
\end{aligned}
\tag{6}
$$

where we have put

$$
A = A(s) = \alpha(s) \wedge \alpha'(s) \wedge \beta(s) \quad \text{and} \quad B = B(s) = \alpha(s) \wedge \beta'(s) \wedge \beta(s).
$$

On the other hand, we note that the vector fields $\alpha(s)$, $\beta(s)$ and $\alpha'(s)$ are mutually orthogonal for all s. Therefore, we can choose another unit vector field $\gamma(s)$ along the base curve α which forms an orthonormal frame in $\mathbb{E}^4$ together with $\alpha(s)$, $\beta(s)$ and $\alpha'(s)$.

Since $\Lambda^3\mathbb{E}^4$ is naturally identified with $\mathbb{E}^4$, we can define the inner product $X_1 \wedge X_2 \wedge X_3$ with X_4 as follows

$$
\ll X_1 \wedge X_2 \wedge X_3, X_4 \gg = \det \begin{pmatrix} X_4 \\ X_1 \\ X_2 \\ X_3 \end{pmatrix},
$$

where the determinant is taken by the 4×4 matrix made up of the components of the vectors X_1, X_2, X_3, X_4 in $\mathbb{E}^4$. Using this inner product, the vector field A is represented by

$$
A = -\alpha \wedge \beta \wedge \alpha' = - \ll \alpha \wedge \beta \wedge \alpha', \gamma(s) \gg \gamma(s) = -\gamma(s)
$$

by considering the orientation and the lengths of vectors. Similarly, we also have

$$
\begin{aligned}
\ll (\alpha \wedge \beta \wedge \alpha')(s), \gamma(s) \gg &= 1, \\
\ll (\alpha \wedge \beta \wedge \gamma)(s), \alpha'(s) \gg &= -1, \\
\ll (\alpha \wedge \alpha' \wedge \gamma)(s), \beta(s) \gg &= 1, \\
\ll (\beta \wedge \alpha' \wedge \gamma)(s), \alpha(s) \gg &= -1
\end{aligned}
\tag{7}
$$

for all s. By virtue of (7), we can obtain the following

$$
\begin{cases}
A = -\gamma, \\
A' = b\beta + \varphi\alpha', \\
A'' = -\varphi\alpha + (b' - u\varphi)\beta + (ub + \varphi')\alpha' + (b^2 + \varphi^2)\gamma, \\
B = b\alpha' - u\gamma, \\
B' = -b\alpha + (b' + u\varphi)\alpha' + (b\varphi - u')\gamma, \\
B'' = -(2b' + u\varphi)\alpha + (u'b - ub' - u^2\varphi - b^2\varphi)\beta \\
\qquad + (b'' + 2u'\varphi + u\varphi' - b - b\varphi^2)\alpha' + (2b'\varphi + u\varphi^2 + b\varphi' - u'')\gamma,
\end{cases}
\tag{8}
$$

which imply that the spherical Gauss map G represented by (4) reduces to

$$
G = \frac{1}{\sqrt{q}}\left((b \sin t)\alpha' - (\cos t + u \sin t)\gamma\right),
\tag{9}
$$

where we have put $b = b(s) = \langle \beta'(s), \gamma(s) \rangle$ and $\varphi = \varphi(s) = \langle \alpha''(s), \gamma(s) \rangle$.

Theorem 1. *Let M be a ruled surface in the sphere* $\mathbb{S}^3$. *Then, M has a harmonic spherical Gauss map if and only if M is totally geodesic in* $\mathbb{S}^3$.

Proof. Suppose that the spherical Gauss map G is harmonic, i.e., $\Delta G = \mathbf{0}$, where $\mathbf{0}$ denotes zero vector. Then, (6) implies

$$\left\{ -(q_s)^2 + \frac{1}{2} q q_{ss} - \frac{1}{2} q(q_t)^2 + \frac{1}{2} q^2 q_{tt} + q^3 \right\} (\cos t A + \sin t B)$$
$$+ \frac{3}{2} q q_s (\cos t A' + \sin t B') - q^2 (\cos t A'' + \sin t B'') + \frac{1}{2} q^2 q_t (-\sin t A + \cos t B) = \mathbf{0}. \tag{10}$$

By the orthogonality of vector fields α, β, α' and γ, putting (8) into (10) gives us

$$\frac{3}{2} q q_s b \sin t - q^2 \varphi \cos t - q^2 \left(2b' + u\varphi \right) \sin t = 0, \tag{11}$$

$$\frac{3}{2} q q_s b \cos t - q^2 \left(b' - u\varphi \right) \cos t - q^2 \left(u'b - ub' - u^2\varphi - b^2\varphi \right) \sin t = 0 \tag{12}$$

as the coefficients of the vectors α and β, respectively. Using the equation for q of (5) and the fact that $q > 0$, (11) and (12) can be expressed as

$$-\varphi \cos^3 t + \left(-3u\varphi - 2b' \right) \cos^2 t \sin t$$
$$+ \left(3u'b - 4ub' - 3u^2\varphi - b^2\varphi \right) \cos t \sin^2 t \tag{13}$$
$$+ \left(3uu'b + b^2 b' - 2u^2 b' - u^3\varphi - ub^2\varphi \right) \sin^3 t = 0$$

and

$$\left(u\varphi - b' \right) \cos^3 t + \left(2u'b + 3u^2\varphi - ub' + b^2\varphi \right) \cos^2 t \sin t$$
$$+ \left(uu'b + 2b^2 b' + 3u^3\varphi + 3ub^2\varphi + u^2 b' \right) \cos t \sin^2 t \tag{14}$$
$$+ \left(u^4\varphi + b^4\varphi + u^3 b' - u^2 u'b + 2u^2 b^2\varphi + ub^2 b' - u'b^3 \right) \sin^3 t = 0,$$

respectively. We easily see that the trigonometric functions of t of (13) and (14) are linearly independent for all t. Therefore, we can see that

$$\varphi = 0 \quad \text{and} \quad b' = 0 \tag{15}$$

by considering the coefficients of the terms containing 'cos^3 t' of (13) and (14), respectively. From the coefficients of the term containing 'cos t sin^2 t' of (13), we get

$$u'b = 0. \tag{16}$$

Suppose that b is a nonzero constant on M. Then, (11) and (16) imply $q_s = 0$. Putting it into (10) yields

$$\left\{ -\frac{1}{2}(q_t)^2 + \frac{1}{2} q q_{tt} + q^2 \right\} (\cos t A + \sin t B)$$
$$- q(\cos t A'' + \sin t B'') + \frac{1}{2} q q_t (-\sin t A + \cos t B) = \mathbf{0}. \tag{17}$$

In this case, the vectors are reduced to

$$
\begin{cases}
A = -\gamma, \\
A'' = ub\alpha' + b^2\gamma, \\
B = b\alpha' - u\gamma, \\
B'' = -b\alpha'.
\end{cases}
\tag{18}
$$

Using (18), we note that Equation (17) can be regarded as the form of the linear combination of two orthogonal vectors α' and γ with trigonometric functions in t as coefficients. By a straightforward computation, we can see that the coefficient of γ of (17) is given by

$$
-2b^2(\cos^5 t + u\cos^4 t \sin t + 2\cos^3 t \sin^2 t + 2u\cos^2 t \sin^3 t + \cos t \sin^4 t + u\sin^5 t)
$$
$$
= -2b^2(\cos t + u\sin t)(\cos^2 t + \sin^2 t)^2
$$
$$
= -2b^2(\cos t + u\sin t) = 0
$$

which implies that $b = 0$, a contradiction to $b \neq 0$. Therefore, the constant b is zero. With the help of (15), we get from (8) that

$$
A' = 0 \quad \text{and} \quad B = uA.
$$

Since the spherical Gauss map $G = -\frac{1}{\sqrt{q}}(\cos t + u\sin t)\gamma$ is a unit normal vector field of the ruled surface M to the unit sphere $\mathbb{S}^3$, it is easily obtained that the shape operator S of M in $\mathbb{S}^3$ vanishes, i.e., M is totally geodesic in $\mathbb{S}^3$.

Conversely, if M is a totally geodesic surface of $\mathbb{S}^3$, i.e., M is a great sphere of $\mathbb{S}^3$, it is not hard to show that the spherical Gauss map of M is harmonic. It completes the proof. $\square$

4. A Ruled Surface in $\mathbb{S}^3$ with a Finite-Type Spherical Gauss Map

In this section, we will investigate a ruled surface M in $\mathbb{S}^3$ parameterized by (2) with a finite-type spherical Gauss map.

Using (6), (8) and (9), the Laplacian ΔG can be put as

$$
\Delta G = -\frac{(q_s)^2}{q^{\frac{7}{2}}}\left((b\sin t)\alpha' - (\cos t + u\sin t)\gamma\right) + \frac{1}{q^{\frac{5}{2}}}P_1(s,t),
$$

where P_1 is a vector field formed with the linear combination of the orthogonal vector fields α, β, α' and γ together with the coefficients of trigonometric functions in t and functions in s. Proceeding by induction, we get

$$
\Delta^m G = a_m \frac{(q_s)^{2m}}{q^{3m+\frac{1}{2}}}\left((b\sin t)\alpha' - (\cos t + u\sin t)\gamma\right) + \frac{1}{q^{3m-\frac{1}{2}}}P_m(s,t)
\tag{19}
$$

for any positive integer m, where P_m is a vector field formed with the linear combination of the orthogonal vector fields α, β, α' and γ together with the coefficients of trigonometric functions in t and functions in s, and a_m is a nonzero constant satisfying $a_m = (3m-1)(\frac{5}{2}-3m)a_{m-1}$ with $a_0 = 1$.

Suppose that the spherical Gauss map G of M is of finite-type. Then, we have

$$
\Delta^k G + c_1\Delta^{k-1}G + c_2\Delta^{k-2}G + \cdots + c_{k-1}\Delta G = 0
\tag{20}
$$

for some constants $c_1, c_2, \ldots, c_{k-1} \in \mathbb{R}$ and a positive integer k. By the orthogonality of the vectors α, β, α' and γ, substituting (19) into (20) gives us the coefficients of α' and γ as follows

$$
(q_s)^{2k}b\sin t = qF_1(s,t)
$$

and

$$(q_s)^{2k}(\cos t + u \sin t) = qF_2(s,t),$$

respectively, from which,

$$\left(2u' \cos t \sin t + w' \sin^2 t\right)^{2k}\left(\cos t + (u+b)\sin t\right)$$
$$= \left(\cos^2 t + 2u \cos t \sin t + w \sin^2 t\right)F(s,t) \tag{21}$$

for some polynomials F_1 and F_2 in 'cos t' and 'sin t' with functions of s as coefficients, where $F(s,t) = F_1(s,t) + F_2(s,t)$.

By the linear independence of the trigonometric functions $\cos^2 t$, $\cos t \sin t$ and $\sin^2 t$, we may put

$$F(s,t) = (q_s)^l R(s,t) = (2u' \cos t \sin t + w' \sin^2 t)^l R(s,t),$$

where l is a non-negative integer less than $2k$ and $R(s,t)$ is some polynomial in 'cos$^{n-k} t \sin^k t$', $k = 0, 1, \ldots, n$, with functions in s as coefficients such that $R(s,t)$ and q_s are relatively prime. That is, $R(s,t)$ is of the form

$$R(s,t) = \sum_{k=0}^{n} \Gamma_k(s) \cos^{n-k} t \sin^k t$$

for some functions Γ_k in s. Here, the degree of $R(s,t)$ is n. Then, (21) becomes

$$\left(2u' \cos t \sin t + w' \sin^2 t\right)^{2k-l}\left(\cos t + (u+b)\sin t\right)$$
$$= \left(\cos^2 t + 2u \cos t \sin t + w \sin^2 t\right)R(s,t). \tag{22}$$

By putting $\theta = \tan t$ in (22), we get

$$\left(\frac{2u'\theta}{(1+\theta^2)} + \frac{w'\theta^2}{(1+\theta^2)}\right)^{2k-l}\left(\frac{1}{\sqrt{1+\theta^2}} + \frac{(u+b)\theta}{\sqrt{1+\theta^2}}\right)$$
$$= \left(\frac{1}{(1+\theta^2)} + \frac{2u\theta}{(1+\theta^2)} + \frac{w\theta^2}{(1+\theta^2)}\right)\frac{1}{(\sqrt{1+\theta^2})^n}\bar{R}(s,\theta),$$

or, equivalently,

$$\left(2u'\theta + w'\theta^2\right)^{2k-l}\left(1+\theta^2\right)^{\frac{n}{2}+1}\left(1 + (u+b)\theta\right)$$
$$= (1 + 2u\theta + w\theta^2)\left(1+\theta^2\right)^{2k-l+\frac{1}{2}}\bar{R}(s,\theta), \tag{23}$$

where $\bar{R}(s,\theta)$ is a polynomial in θ with functions in s as coefficients such that

$$\bar{R}(s,\theta) = \left(\sqrt{1+\theta^2}\right)^n R(s,t(\theta)).$$

We note that two polynomials $(2u'\theta + w'\theta^2)$ and $\bar{R}(s,\theta)$ are relatively prime, where the former one is obtained from q_s.

Now, we will deal with possible cases derived from (23). Considering the degree of (23) with respect to θ and the linear independence of $(2u'\theta + w'\theta^2)$ and $\bar{R}(s,\theta)$, we can put

$$\left(2u'\theta + w'\theta^2\right)^{2k-l} = \lambda(s)(1 + 2u\theta + w\theta^2)\left(1+\theta^2\right)^{2k-l-1},$$

or,

$$\left(2u'\theta + w'\theta^2\right)^{2k-l}(1+\theta^2) = \lambda(s)(1 + 2u\theta + w\theta^2)\left(1+\theta^2\right)^{2k-l} \tag{24}$$

for some function λ in s.

Recall that (24) is a polynomial in θ. So, by comparing the smallest power of both sides of (24) with respect to θ, we can see that '$2k - l$' must be zero. Therefore, (24) becomes of the form

$$(1 + \theta^2) = \lambda(s)(1 + 2u\theta + w\theta^2).$$

It follows that

$$u = 0 \quad \text{and} \quad w = 1,$$

from which, we get the function q is constant with value 1 and the metric tensor g of M is given by

$$g = \begin{pmatrix} 1 & 0 \\ 0 & 1 \end{pmatrix},$$

from which, we see that M is flat in $\mathbb{E}^4$. It also gives us $\Delta x = 2x$ and hence M is minimal in $\mathbb{S}^3$. Therefore, M is one of the isoparametric surfaces in $\mathbb{S}^3$, which is the Clifford torus $\mathbb{S}^1(1/\sqrt{2}) \times \mathbb{S}^1(1/\sqrt{2})$.

Together with Theorem 1, we have

Theorem 2. *Let M be a complete ruled surface in the sphere $\mathbb{S}^3$ with finite-type spherical Gauss map. Then, M is either the Clifford torus $\mathbb{S}^1(1/\sqrt{2}) \times \mathbb{S}^1(1/\sqrt{2})$ or a totally geodesic surface in $\mathbb{S}^3$.*

Corollary 1. *Let M be a ruled surface in the sphere $\mathbb{S}^3$. If the spherical Gauss map G of M is of finite-type, then both M and G are of 1-type. In particular, $\Delta x = 2x$ and either $\Delta G = \boldsymbol{0}$ or $\Delta G = 2G$.*

5. Ruled Surfaces in $\mathbb{S}^3$ with Pointwise 1-Type Spherical Gauss Maps of the First Kind

In this section, we will study a ruled surface in $\mathbb{S}^3$ with pointwise 1-type spherical Gauss map G of the first kind, i.e., $\Delta G = fG$ for some nonzero smooth function f. Let M be a ruled surface in the sphere $\mathbb{S}^3$ ($\subset \mathbb{E}^4$) parameterized by (2). Then, using (6), equation $\Delta G = fG$ gives

$$\left\{ -(q_s)^2 + \frac{1}{2}qq_{ss} - \frac{1}{2}q(q_t)^2 + \frac{1}{2}q^2 q_{tt} + (1-f)q^3 \right\}(\cos tA + \sin tB)$$

$$+ \frac{3}{2}qq_s(\cos tA' + \sin tB') - q^2(\cos tA'' + \sin tB'') + \frac{1}{2}q^2 q_t(-\sin tA + \cos tB) = 0. \tag{25}$$

With the help of (8), by comparing two equations, (10) and (25), we can see that the coefficients of the vectors α and β of (25) coincide with those of α and β of (10). Therefore, we obtain (13) and (14), or, equivalently, we have

$$\varphi = 0, \quad b' = 0 \quad \text{and} \quad u'b = 0.$$

Similarly as we did to the constant b in Section 3, we will show the constant b is nonzero and hence u is a constant. Suppose that $b = 0$ on M. $\Delta G = fG$ with $b = 0$ gives

$$\left\{ -(q_s)^2 + \frac{1}{2}qq_{ss} - \frac{1}{2}q(q_t)^2 + \frac{1}{2}q^2 q_{tt} + (1-f)q^3 \right\}(-\cos t - u\sin t)$$

$$+ \frac{3}{2}qq_s(-u'\sin t) - q^2(-u''\sin t) + \frac{1}{2}q^2 q_t(\sin t - u\cos t) = 0,$$

from which,

$$f(\cos t + u\sin t)^7 = 0.$$

It implies that f is vanishing. It is a contradiction and thus we conclude that b is nonzero. Then, we have $q_s = 0$. With the help of (18), (25) is reduced to

$$\left\{ -\frac{1}{2}(q_t)^2 + \frac{1}{2}qq_{tt} + (1-f)q^2 \right\}(\cos tA + \sin tB)$$
$$-q(\cos tA'' + \sin tB'') + \frac{1}{2}qq_t(-\sin tA + \cos tB) = 0$$

which provides us with

$$fq^2 \sin t = \left\{ -\frac{1}{2}(q_t)^2 + \frac{1}{2}qq_{tt} + q^2 + q \right\}\sin t + \left(\frac{1}{2}qq_t - uq\right)\cos t$$

as the coefficients of the vector α'.

We note that $w = u^2 + b^2$. By a straightforward computation, we get

$$f = \frac{2b^2}{q^2}.$$

Consequently, if a ruled surface M has pointwise 1-type spherical Gauss map of the first kind, we see that the constant b is nonzero and $\varphi = 0$, that is, the curves α and β satisfy

$$\alpha' \wedge \beta' \neq 0 \quad \text{and} \quad \alpha'' \wedge \alpha \wedge \beta = 0 \tag{26}$$

for all s. Now, we consider the curve $\delta(s)$ on the sphere $\mathbb{S}^3(\sqrt{\frac{u^2+1}{b^2}})$ given by

$$\delta(s) = -\frac{u}{b}\alpha(s) + \frac{1}{b}\beta(s).$$

We note that the curve $\delta(s)$ is an integral curve of γ, that is, $\delta' = \gamma$. Then, we can easily show that the spherical Gauss map G of a ruled surface M in $\mathbb{S}^3$ parameterized by

$$M : x(s,t) = \alpha(s)\cos t + \beta(s)\sin t$$
$$= \left(\cos t + u\sin t\right)\alpha(s) + b\sin t\delta(s) \tag{27}$$

is of pointwise 1-type of the first kind. Indeed, it follows that

$$\Delta G = \left(\frac{2b^2}{q^2}\right)G.$$

Therefore, we have

Theorem 3. *Let M be a ruled surface in the unit sphere $\mathbb{S}^3$. If M has pointwise 1-type spherical Gauss map of the first kind, then M is part of the ruled surface in $\mathbb{S}^3$ parameterized by (27) satisfying (26).*

Example 1. *The curves $\alpha(s)$ and $\delta(s)$, given by*

$$\alpha(s) = \left(\frac{1}{\sqrt{2}}\cos s, \frac{1}{\sqrt{2}}\sin s, \frac{1}{\sqrt{2}}\cos s, \frac{1}{\sqrt{2}}\sin s\right)$$

and

$$\delta(s) = \left(\frac{1}{2\sqrt{2}}\sin(2s), -\frac{1}{2\sqrt{2}}\cos(2s), -\frac{1}{2\sqrt{2}}\sin(2s), \frac{1}{2\sqrt{2}}\cos(2s)\right)$$

are unit speed curves on the sphere $\mathbb{S}^3$ and the sphere $\mathbb{S}^3(\frac{1}{2})$, respectively. In this case, it is clear that $u = 0$ and $b = 2$. Then, the ruled surface M in the sphere $\mathbb{S}^3$ defined by

$$M : x(s,t) = \alpha(s) \cos t + 2\delta(s) \sin t$$
$$= \frac{1}{\sqrt{2}} \Big(\cos t \cos s + \sin t \sin(2s), \cos t \sin s - \sin t \cos(2s),$$
$$\cos t \cos s - \sin t \sin(2s), \cos t \sin s + \sin t \cos(2s) \Big)$$

has pointwise 1-type spherical Gauss map G of the first kind

$$\Delta G = \frac{8}{\Big(\cos^2 t + 4 \sin^2 t \Big)^2} G.$$

6. Ruled Surfaces in $\mathbb{S}^3$ with Pointwise 1-Type Spherical Gauss Maps of the Second Kind

In this section, we will investigate a ruled surface M in $\mathbb{S}^3$ parameterized by (2) with a pointwise 1-type spherical Gauss map of the second kind, that is, the spherical Gauss map G of M satisfies

$$\Delta G = f(G + \mathbf{C})$$

for some nonzero function f of s and t and a non-zero constant vector $\mathbf{C}$. If we consider a non-empty open subset $U = \{(s,t) \in I \times J \mid f(s,t) \neq 0\}$, then we can put

$$\mathbf{C} = \frac{\Delta G - fG}{f} \tag{28}$$

which yields that

$$f(\Delta G - fG)_t = f_t(\Delta G - fG) \tag{29}$$

on U.

Now, we consider the open subset $U_0 = \{(s,t) \in U \mid f_t(s,t) \neq 0\}$ and suppose that U_0 is non-empty. With the help of (8) and (25), we can get from (29),

$$f(q^{-\frac{7}{2}}P)_t = f_t(q^{-\frac{7}{2}}P) \quad \text{and} \quad f(q^{-\frac{7}{2}}Q)_t = f_t(q^{-\frac{7}{2}}Q), \tag{30}$$

or, equivalently,

$$f\Big(-\frac{7}{2}q_t P + qP_t \Big) = qf_t P \quad \text{and} \quad f\Big(-\frac{7}{2}q_t Q + qQ_t \Big) = qf_t Q$$

as the coefficients of the vectors α and β of (29), respectively, where we have put

$$P(s,t) = \frac{3}{2}qq_s b \sin t - q^2 \varphi \cos t - q^2 \Big(u\varphi + 2b' \Big) \sin t \tag{31}$$

and

$$Q(s,t) = \frac{3}{2}qq_s b \cos t + q^2 \Big(u\varphi - b' \Big) \cos t - q^2 \Big(u'b - ub' - u^2\varphi - b^2\varphi \Big) \sin t. \tag{32}$$

Now, we will consider a few lemmas to reach a conclusion for this section.

Lemma 1. *Let M be a ruled surface in the unit sphere $\mathbb{S}^3$ parameterized by* (2) *with a pointwise 1-type spherical Gauss map of the second kind. If $U_0 = \{(s,t) \in I \times J \mid f_t(s,t) \neq 0\}$ $(\subset U)$ is non-empty, then*

$$\alpha' \wedge \beta' = 0 \quad \text{on} \quad U_0.$$

Proof. We suppose that the function $b(s)$ is non-vanishing on some open set U_1 of U_0. We first consider the case that at least one of two equations $P(s,t)$ and $Q(s,t)$ is vanishing on some subset of U_1, say $P(s,t) = 0$. Then, we can easily show that

$$\varphi = 0 \quad \text{and} \quad b' = 0 \tag{33}$$

by considering the linear independence of the trigonometric functions of (31). Since b is a nonzero constant, (31) and (33) imply $q_s = 0$. Thus, the function $Q(s,t)$ of (32) has to be identically zero on that subset U_1. Similarly, if $Q(s,t) = 0$, we can derive $P(s,t) = 0$. Therefore, we suppose that both $\Phi(s,t)$ and $\Psi(s,t)$ are identically zero on U_1. In this case, Equation (29) can be put as

$$\left(f_t q^{-\frac{5}{2}} \Lambda_1\right) \alpha' + \left(f_t q^{-\frac{5}{2}} \Lambda_2\right) \gamma = \left(f(q^{-\frac{5}{2}} \Lambda_1)_t\right) \alpha' + \left(f(q^{-\frac{5}{2}} \Lambda_2)_t\right) \gamma$$

which yields that

$$\frac{f_t}{f} = \frac{(q^{-\frac{5}{2}} \Lambda_1)_t}{q^{-\frac{5}{2}} \Lambda_1} = \frac{(q^{-\frac{5}{2}} \Lambda_2)_t}{q^{-\frac{5}{2}} \Lambda_2} \tag{34}$$

by comparing the coefficients of two orthogonal vectors α' and γ, where we have put

$$\Lambda_1(s,t) = \left\{ -\frac{1}{2}(q_t)^2 + \frac{1}{2}qq_{tt} + (1-f)q_2 \right\} b \sin t - qb(u \cos t - \sin t) + \frac{1}{2}qq_t b \cos t \tag{35}$$

and

$$\Lambda_2(s,t) = \left\{ \frac{1}{2}(q_t)^2 - \frac{1}{2}qq_{tt} - (1-f)q_2 \right\} (\cos t + u \sin t)$$
$$- qb^2 \cos t + \frac{1}{2}qq_t (\sin t - u \cos t). \tag{36}$$

By taking the integration to (34) with respect to t, we see that the function f takes the form

$$f = y_1(s)(q^{-\frac{5}{2}} \Lambda_1) = y_2(s)(q^{-\frac{5}{2}} \Lambda_2) \tag{37}$$

for some non-vanishing functions y_1 and y_2 of s. If we put (35) and (36) into (37), then we can obtain the formulas for f as

$$f = \frac{2b^3 y_1(s) \sin t}{q^2 \left(q^{\frac{1}{2}} + by_1(s) \sin t\right)} = \frac{-2b^2 y_2(s)(\cos t + u \sin t)}{q^2 \left(q^{\frac{1}{2}} - y_2(s)(\cos t + u \sin t)\right)}. \tag{38}$$

Comparing the last two equations in (38), we get

$$\left(y_2 \cos t + (by_1 + uy_2) \sin t\right) q^{\frac{1}{2}} = 0$$

which implies

$$y_2(s) = 0 \quad \text{and} \quad y_1(s) = 0$$

because of $q \neq 0$, but it contradicts $f \neq 0$. Consequently, this case never occurs. Therefore, we may assume that both $P(s,t)$ and $Q(s,t)$ are both non-vanishing on $U_1 (\subset U_0)$. Then, equations of (30) give

$$\frac{f_t}{f} = \frac{(q^{-\frac{7}{2}}P)_t}{(q^{-\frac{7}{2}}P)} = \frac{(q^{-\frac{7}{2}}Q)_t}{(q^{-\frac{7}{2}}Q)} \tag{39}$$

on U_1 and thus the function f is of the form

$$f = g_1(s)q^{-\frac{7}{2}}P(s,t) = g_2(s)q^{-\frac{7}{2}}Q(s,t) \tag{40}$$

which implies

$$\begin{aligned}
g_1(s) &\Big\{ \frac{3}{2}b(2u'\cos t\sin^2 t + w'\sin^3 t) - \varphi(\cos^3 t + 2u\cos^2 t\sin t + w\cos t\sin^2 t) \\
&\quad - \Big(u\varphi - 2b' \Big)(\cos^2 t\sin t + 2u\cos t\sin^2 t + w\sin^3 t) \Big\} \\
= g_2(s) &\Big\{ \frac{3}{2}b(2u'\cos^2 t\sin t + w'\cos t\sin^2 t) \\
&\quad + \Big(u\varphi - b' \Big)(\cos^3 t + 2u\cos^2 t\sin t + w\cos t\sin^2 t) \\
&\quad - \Big(u'b - ub' - u^2\varphi - b^2\varphi \Big)(\cos^2 t\sin t + 2u\cos t\sin^2 t + w\sin^3 t) \Big\}
\end{aligned} \tag{41}$$

for some non-vanishing functions g_1 and g_2 of s on U_1 because of $q > 0$. By the linear independence of trigonometric functions $\cos^{3-k} t \sin^k t$ of (41) for $k = 0, \ldots, 3$, we have

$$g_1\varphi = g_2\Big(b' - u\varphi \Big), \tag{42}$$

$$g_1\Big(-3u\varphi - 2b' \Big) = g_2\Big(2u'b + 3u^2\varphi - ub' + b^2\varphi \Big), \tag{43}$$

$$g_1\Big(3u'b - 4ub' - 3u^2\varphi - b^2\varphi \Big) = g_2\Big(uu'b + 2b^2b' + 3u^3\varphi + 3ub^2\varphi + u^2b' \Big) \tag{44}$$

and

$$\begin{aligned}
g_1\Big(3uu'b + b^2b' - 2u^2b' - u^3\varphi - ub^2\varphi \Big) \\
= g_2\Big(u^4\varphi + b^4\varphi + u^3b' - u^2u'b + 2u^2b^2\varphi + ub^2b' - u'b^3 \Big)
\end{aligned} \tag{45}$$

as the coefficients of terms containing 'cos$^3 t$', 'cos$^2 t \sin t$', 'cos$t \sin^2 t$' and 'sin$^3 t$', respectively. Substituting (42) into (43), we get

$$-2b'g_1 = g_2\Big(2ub' + 2u'b + b^2\varphi \Big) \tag{46}$$

which implies

$$u'bg_1 = g_2\Big(-uu'b + b^2b' \Big) \tag{47}$$

with the aid of (42) and (44). Finally, putting (42), (46) and (47) into (45) allows us to have

$$\frac{3}{2}b^4\varphi g_2 = 0$$

and hence $\varphi = 0$ because b and g_2 are non-vanishing on U_1. From (42) and (43), we can see that

$$b' = 0 \quad \text{and} \quad u' = 0,$$

or, equivalently,

$$q_s = 0 \quad \text{on} \quad U_1.$$

Since $\varphi = 0$ and $q_s = 0$ on U_1, the non-vanishing function $P(s,t)$ of (31) on U_0 becomes identically zero on $U_1 \subset U_0$, a contradiction. Therefore, we conclude that the set U_1 is empty, which means that $\beta' = u\alpha'$ on U_0 as we desired. $\square$

Now, we will examine the set $x(U_0)$ of $\mathbb{S}^3$. In Lemma 1, we showed that $b = \langle \beta', \gamma \rangle = 0$ on U_0. Then, we have

$$q = (\cos t + u(s) \sin t)^2$$

and the spherical Gauss map G of (9) is given by

$$G = -\gamma. \tag{48}$$

From (41), we see that

$$\Big((g_1 + ug_2)\varphi\Big)(s) = 0 \quad \text{on} \quad U_0. \tag{49}$$

If $\varphi = 0$ on some subset U_2 of U_0 with $int(U_2) \neq \varnothing$, then

$$G' = (-\gamma)' = \varphi\alpha' = 0$$

which means that the spherical Gauss map G is constant and thus $\Delta G = 0$ on that subset. Since the spherical Gauss map is of pointwise 1-type of the second kind and $\mathbf{C}$ is a constant vector, $G = -\mathbf{C}$ globally.

Now, we suppose that the function φ is non-vanishing on U_0. From (49), we see that $(g_1 + ug_2) \equiv 0$ on U_0 and then, the function f of (40) is simplified as

$$f(s,t) = -\frac{\varphi(s)g_1(s)}{(\cos t + u(s) \sin t)^2}, \tag{50}$$

so equation $\Delta G = f(G + \mathbf{C})$ can be expressed as follows

$$\frac{u' \sin t}{(\cos t + u \sin t)^3}\gamma' - \frac{1}{(\cos t + u \sin t)^2}\gamma'' = \frac{\varphi g_1}{(\cos t + u \sin t)^2}(-\gamma + \mathbf{C}).$$

With the help of (8), it follows that

$$u'\varphi \sin t\alpha' + (\cos t + u \sin t)(\varphi\alpha + u\varphi\beta - \varphi'\alpha' - \varphi^2\gamma) \tag{51}$$
$$= \varphi g_1(\cos t + u \sin t)(\gamma - \mathbf{C})$$

which guarantees that

$$\varphi\alpha + u\varphi\beta - \varphi'\alpha' - \varphi^2\gamma = \varphi g_1(\gamma - \mathbf{C}) \tag{52}$$

by considering the terms containing '$\cos t$'. Thus, the constant vector $\mathbf{C}$ can be put

$$\mathbf{C} = -\frac{1}{g_1}\alpha - \frac{u}{g_1}\beta + \frac{\varphi'}{\varphi g_1}\alpha' + \left(\frac{\varphi}{g_1} + 1\right)\gamma, \tag{53}$$

from which,

$$0 = -\left(\left(\frac{1}{g_1}\right)' + \frac{\varphi'}{\varphi g_1}\right)\alpha - \left(\left(\frac{u}{g_1}\right)' + \frac{u\varphi'}{\varphi g_1}\right)\beta$$
$$+ \left(-\frac{1}{g_1} - \frac{u^2}{g_1} + \left(\frac{\varphi'}{\varphi g_1}\right)' - \frac{\varphi^2}{g_1} - \varphi\right)\alpha' + \left(\frac{\varphi'}{g_1} + \left(\frac{\varphi}{g_1}\right)'\right)\gamma.$$

By (51) and (52), we note that $u' = 0$ on U_0. Thus, the above equation provides us with the following equations

$$\begin{cases} \left(\dfrac{1}{g_1}\right)' = -\dfrac{\varphi'}{\varphi g_1}, \\[2mm] \left(\dfrac{\varphi'}{\varphi g_1}\right)' = \dfrac{1 + u^2 + \varphi^2}{g_1} + \varphi, \\[2mm] \left(\dfrac{\varphi}{g_1}\right)' = -\dfrac{\varphi'}{g_1} \end{cases} \tag{54}$$

as the coefficients of the orthogonal vectors. Comparing the first and the third equations of (54), we can obtain

$$\varphi' = 0$$

which yields that

$$g_1' = 0 \quad \text{and} \quad \left(\frac{1 + u^2 + \varphi^2}{g_1} + \varphi\right) = 0 \quad \text{on} \quad U_0.$$

Therefore, we see that the function φ is nonzero constant on U_0. The functions u and g_1 are also constant on U_0, so is the function g_2 by virtue of (49). Since $g_1 \varphi = -(1 + u^2 + \varphi^2)$, we have

$$f = \frac{1 + u^2 + \varphi^2}{(\cos t + u \sin t)^2}$$

and

$$\mathbf{C} = \frac{1}{1 + u^2 + \varphi^2} \left(\varphi \alpha + u \varphi \beta + (1 + u^2)\gamma\right)$$

from (50) and (53), respectively.

According to the results so far, we are ready to construct a ruled surface M in $\mathbb{S}^3$ with a pointwise 1-type spherical Gauss map of the second kind which is not totally geodesic, i.e., $\Delta G \neq 0$: As we saw in Lemma 1, if a ruled surface M in $\mathbb{S}^3$ has a pointwise 1-type spherical Gauss map G of the second kind, then $\alpha' \wedge \beta' = 0$ on M. Furthermore, we showed that $q_s = 0$ on M and hence

$$\beta(s) = u \alpha(s) + \mathbf{N},$$

where $\mathbf{N}$ is some constant vector satisfying

$$\langle \alpha, \mathbf{N} \rangle = -u \quad \text{and} \quad \langle \mathbf{N}, \mathbf{N} \rangle = 1 + u^2.$$

Since the function φ is nonzero constant, we can see that the vector field α'' given by

$$\alpha'' = -\alpha - u\beta + \varphi\gamma$$

has the constant length $\sqrt{1 + u^2 + \varphi^2}$. Thus, we can naturally define a ruled surface M in $\mathbb{S}^3$ ($\subset \mathbb{E}^4$) parameterized by

$$M : x(s,t) = \cos t\, \alpha(s) + \sin t\, \beta(s)$$
$$= (\cos t + u \sin t)\alpha(s) + \sin t\, \mathbf{N} \tag{55}$$

which has pointwise 1-type spherical Gauss map G of the second kind, that is,

$$\Delta G = \frac{1}{(\cos t + u \sin t)^2} \gamma''$$
$$= \frac{1}{(\cos t + u \sin t)^2} \left(\varphi(1 + u^2)\alpha + u\varphi \mathbf{N} - \varphi^2 \gamma\right)$$
$$= f(G + \mathbf{C}),$$

where we have put

$$f = \frac{1 + u^2 + \varphi^2}{(\cos t + u \sin t)^2}$$

and

$$\mathbf{C} = \frac{1}{1 + u^2 + \varphi^2} \Big(\varphi(1 + u^2)\alpha + u\varphi\mathbf{N} + (1 + u^2)\gamma \Big),$$

respectively.

Meanwhile, we note that the function φ is constant on U_0. By continuity, we see that either $\Delta G = 0$ on U_0, or it does not. This means that either $G = -\mathbf{C}$ on U_0 or $x(U_0)$ is an open part of a ruled surface parameterized by (55).

Now, we consider $W = \{(s,t) \in U \mid f_t(s,t) = 0\}$, the complement of U_0, and let $W_0 = int(W)$. Then, we will show that if W_0 is non-empty, the constant vector $\mathbf{C} = \mathbf{0}$ on W_0, which implies that W_0 must be empty. Therefore, we have

Lemma 2. *Let M be a ruled surface parameterized by* (2) *in the unit sphere* $\mathbb{S}^3$. *If the spherical Gauss map G of M is of pointwise 1-type of the second kind, i.e.,* $\Delta G = f(G + \mathbf{C})$ *for some non-zero function f and a non-zero constant vector* $\mathbf{C}$, *then we may assume that the function* f_t, *the partial derivative of f with respect to t, is non-vanishing on* $U = \{(s,t) \in I \times J \mid f(s,t) \neq 0\}$, *that is,* $W_0 = \emptyset$.

Proof. We suppose that W_0 is non-empty. From (30), we have

$$(q^{-\frac{7}{2}}P)_t = (q^{-\frac{7}{2}}Q)_t = 0,$$

or, equivalently,

$$\frac{7}{2}q_t P = qP_t \quad \text{and} \quad \frac{7}{2}q_t Q = qQ_t \tag{56}$$

on W_0. By a straightforward computation, $\frac{7}{2}q_t P = qP_t$ of (56) implies

$$\begin{aligned}
\frac{15}{4}bq_s q_t \sin t =& \frac{3}{2}bqq_{st}\sin t + \frac{3}{2}bqq_s\cos t + \frac{3}{2}\varphi qq_t\cos t + \varphi q^2 \sin t \\
&+ \frac{3}{2}(u\varphi + 2b')qq_t\sin t - q^2(u\varphi + 2b')\cos t.
\end{aligned} \tag{57}$$

We note that

$$\begin{cases}
q = \cos^2 t + 2u\cos t\sin t + w\sin^2 t, \\
q_s = 2u'\cos t\sin t + w'\sin^2 t, \\
q_t = 2u\cos^2 t + 2(w-1)\cos t\sin t - 2u\sin^2 t, \\
q_{st} = 2u'\cos^2 t + 2w'\cos t\sin t - 2u'\sin^2 t.
\end{cases} \tag{58}$$

Therefore, we can see that Equation (57) is a polynomial in $\cos^{5-k} t \sin^k t$, $k = 0, 1, \ldots, 5$, with functions of s as coefficients. By considering the linear independence of the trigonometric functions, we get

$$u\varphi = b' \tag{59}$$

as the coefficients of terms containing '$\cos^5 t$'. Thus, the function Q of (32) becomes

$$Q(s,t) = \frac{3}{2}bqq_s\cos t - q^2(u'b - ub' - u^2\varphi - b^2\varphi)\sin t$$

and then $\frac{7}{2}q_t Q = qQ_t$ of (56) provides

$$\frac{15}{4}bq_s q_t \cos t = \frac{3}{2}bqq_{st}\cos t - \frac{3}{2}bqq_s \sin t + \frac{3}{2}(u'b - ub' - u^2\varphi - b^2\varphi)qq_t \sin t \qquad (60)$$
$$- q^2(u'b - ub' - u^2\varphi - b^2\varphi)\cos t.$$

Similarly, using (58), we obtain

$$2u'b + 2u^2\varphi + b^2\varphi = 0 \qquad (61)$$

and

$$u(u'b - u^2\varphi) = 0 \qquad (62)$$

as the coefficients of the terms containing '$\cos^5 t$' and '$\sin^5 t$' of (60), respectively. If $u \neq 0$ on some open subset W_1 of W_0, then we have

$$u'b = u^2\varphi$$

which helps (61) lead to

$$(4u^2 + b^2)\varphi = 0,$$

or,

$$\varphi = 0 \quad \text{on} \quad W_1 \qquad (63)$$

because of $u \neq 0$. Since $\varphi = 0$, (59) and (62) yield that

$$b' = u'b = 0 \quad \text{on} \quad W_1. \qquad (64)$$

If $b = 0$, the function $q = (\cos t + u\sin t)^2$ and the spherical Gauss map G is given by

$$G = A = -\gamma$$

that is constant because of (8) and (63). In this case, we can easily show that the shape operator on W_1 is identically zero, which means that $x(W_1)$ is totally geodesic in $\mathbb{S}^3$.

Now, we may assume that $b \neq 0$ on W_1. It follows that $u' = 0$ of (64) and hence, by continuity, u and b are nonzero constant on W_0, which tells us that

$$q_s = 0 \quad \text{on} \quad W_0.$$

If $u = 0$ on W_0, it is obvious that $b' = 0$ of (59) and hence $q_s = 0$ on W_0. But, in the course of proving $q_s = 0$, we showed that $\varphi = 0$ on W_0. For the case of $u \neq 0$ on W_0, we have (63). If $u = 0$ on W_0, (61) yields that $\varphi = 0$ on W_0. Using these results on W_0, we have

$$\Delta G = q^{-\frac{5}{2}}\left\{ \left(-\frac{1}{2}(q_t)^2 + \frac{1}{2}qq_{tt} + q^2 \right)(\cos t A + \sin t B) \right.$$
$$\left. - q(\cos t A'' + \sin t B'') + \frac{1}{2}qq_t(-\sin t A + \cos t B) \right\},$$

$$\begin{cases} A = -\gamma, \\ A'' = uba' + b^2\gamma \end{cases} \text{and} \begin{cases} B = ba' - u\gamma, \\ B'' = -ba'. \end{cases}$$

By a straightforward computation, we can obtain

$$\Delta G = \frac{2b^2}{q^2}G$$

which means that the spherical Gauss map G defined on W_0 cannot be of pointwise 1-type of the second kind. $\quad\square$

By Lemma 2, we conclude that $U = U_0$. Then, according to the value of the constant function φ, that is, zero or not, it follows that either $G = -\mathbf{C}$ on U or $x(U)$ is an open part of a ruled surface parameterized by (55). On the other hand, Theorem 1 shows that if the interior of the set $\{p \in M | f(p) = 0\}$ of M is non-empty, then it is an open part of a totally geodesic surface in $\mathbb{S}^3$. In fact, a totally geodesic surface of $\mathbb{S}^3$ has a constant spherical Gauss map. And, we can easily show that the function φ defined on a totally geodesic surface of $\mathbb{S}^3$ is identically zero for all s.

Lemma 3. *Let M be a ruled surface in $\mathbb{S}^3$ parameterized by (2) with pointwise 1-type spherical Gauss map of the second kind. Then, the function $\varphi(s) = \langle \alpha''(s), \gamma(s) \rangle$ defined on M is constant for all s.*

By continuity of φ, we can see that if a ruled surface M of $\mathbb{S}^3$ has the spherical Gauss map of pointwise 1-type of the second kind, then we may assume that either M is part of the ruled surface parameterized by (55) or $\Delta G = 0$ on M, given by $G = -\mathbf{C}$. Therefore, we have

Theorem 4. *Let M be a ruled surface in the unit sphere $\mathbb{S}^3$ with a pointwise 1-type spherical Gauss map of the second kind. Then, M is an open part of either the ruled surface parameterized by (55) or a totally geodesic surface.*

Example 2. *Let us consider a unit speed curve α on $\mathbb{S}^3$ and a constant vector $\mathbf{N}$ in $\mathbb{E}^4$ given by*

$$\alpha(s) = \left(\frac{1}{\sqrt{2}} \cos \sqrt{2}s, \frac{1}{\sqrt{2}} \sin \sqrt{2}s, \frac{1}{\sqrt{2}}, 0 \right)$$

and

$$\mathbf{N} = \left(0, 0, 1, -\frac{1}{\sqrt{2}} \right).$$

Then, we get $\langle \alpha, \mathbf{N} \rangle = \frac{1}{\sqrt{2}}$ for all s. By the same argument to get (55), we have

$$\beta(s) = \left(-\frac{1}{2} \cos \sqrt{2}s, -\frac{1}{2} \sin \sqrt{2}s, \frac{1}{2}, -\frac{1}{\sqrt{2}} \right),$$

$$\gamma(s) = \left(-\frac{1}{2} \cos \sqrt{2}s, -\frac{1}{2} \sin \sqrt{2}s, \frac{1}{2}, \frac{1}{\sqrt{2}} \right).$$

Therefore, the ruled surface M on $\mathbb{S}^3$ parameterized by

$$M : x(s,t) = \cos t\alpha(s) + \sin t\beta(s)$$
$$= \left(\frac{1}{\sqrt{2}}(\cos t - \frac{1}{\sqrt{2}} \sin t) \cos \sqrt{2}s, \frac{1}{\sqrt{2}}(\cos t - \frac{1}{\sqrt{2}} \sin t) \sin \sqrt{2}s, \right.$$
$$\left. \frac{1}{\sqrt{2}} \cos t + \frac{1}{2} \sin t, -\frac{1}{\sqrt{2}} \sin t \right),$$

has the spherical Gauss map G of the form

$$G = -\gamma = \left(\frac{1}{2} \cos \sqrt{2}s, \frac{1}{2} \sin \sqrt{2}s, -\frac{1}{2}, -\frac{1}{\sqrt{2}} \right),$$

which satisfies

$$\Delta G = \frac{2}{\left(\cos t - \frac{1}{\sqrt{2}} \sin t \right)^2} \left(G + (0, 0, \frac{1}{2}, \frac{1}{\sqrt{2}}) \right).$$

Author Contributions: S.M.J. and Y.H.K. set up the problem and computed the details and J.Q. checked and polished the draft.

Funding: The first author was supported by the National Research Foundation of Korea (NRF) grant funded by the Korea government (MSIT) (2019R1C1C1006370).

Conflicts of Interest: The authors declare no conflict of interest.

References

1. Takahashi, T. Minimal immersions of Riemannian manifolds. *J. Math. Soc. Jpn.* **1966**, *18*, 380–385. [CrossRef]
2. Chen, B.-Y. *Finite-Type Submanifolds and Generalizations*; Instituto "Guido Castelnuovo": Rome, Italy, 1985.
3. Chen, B.-Y. Surfaces of finite-type in Euclidean 3-space. *Bull. Soc. Math. Belg.* **1987**, *39*, 243–254.
4. Chen, B.-Y.; Dillen, F.; Verstraelen, L.; Vrancken, L. Ruled surfaces of finite-type. *Bull. Aust. Math. Soc.* **1990**, *42*, 447–453. [CrossRef]
5. Dillen, F. Ruled submanifolds of finite-type. *Proc. Am. Math. Soc.* **1992**, *114*, 795–798. [CrossRef]
6. Kim, Y.H.; Yoon, D.W. Classification of ruled surfaces in Minkowski 3-spaces. *J. Geom. Phys.* **2004**, *49*, 89–100. [CrossRef]
7. Kim, Y.H.; Yoon, D.W. On non-developable ruled surfaces in Lorentz-Minkowski 3-spaces. *Taiwan. J. Math.* **2007**, *11*, 197–214. [CrossRef]
8. Hasanis, T.; Vlachos, T. A classification of ruled surfaces of finite type in $\mathbb{S}^3$. *J. Geom.* **1994**, *50*, 84–94. [CrossRef]
9. Chen, B.-Y.; Piccinni, P. Submanifolds with finite-type Gauss map. *Bull. Aust. Math. Soc.* **1987**, *35*, 161–186. [CrossRef]
10. Baikoussis, C. Ruled submanifolds with finite-type Gauss map. *J. Geom.* **1994**, *49*, 42–45. [CrossRef]
11. Baikoussis, C.; Blair, D.E. On the Gauss map of ruled surfaces. *Glasg. Math. J.* **1992**, *34*, 355–359 [CrossRef]
12. Baikoussis, C.; Chen, B.-Y.; Verstraelen, L. Ruled surfaces and tubes with finite-type Gauss map. *Tokyo J. Math.* **1993**, *16*, 341–348. [CrossRef]
13. Jang, C. Surfaces with 1-tpye Gauss map. *Kodai Math. J.* **1996**, *19*, 388–394. [CrossRef]
14. Kim, D.-S.; Kim, Y.H.; Jung, S.M. Ruled submanifolds with harmonic Gauss map. *Taiwan. J. Math.* **2014**, *18*, 53–76. [CrossRef]
15. Kim, D.-S.; Kim, Y.H.; Jung, S.M. Some classifications of ruled submanifolds in Minkowski space and their Gauss map. *Taiwan. J. Math.* **2014**, *18*, 1021–1040. [CrossRef]
16. Kim, D.-S.; Kim, Y.H.; Yoon, D.W. Extended B-scrolls and their Gauss maps. *Indian J. Pure Appl. Math.* **2002**, *33*, 1031–1040.
17. Kim, Y.H.; Yoon, D.W. Ruled surfaces with finite type Gauss map in Minkowski spaces. *Soochow J. Math.* **2000**, *26*, 85–96.
18. Chen, B.-Y.; Choi, M.; Kim, Y.H. Surfaces of revolution with pointwise 1-type Gauss map. *J. Korean Math. Soc.* **2005**, *42*, 447–455. [CrossRef]
19. Choi, M.; Kim, Y.H. Characterization of the helicoid as ruled surfaces with pointwise 1-type Gauss map. *Bull. Korean Math. Soc.* **2001**, *38*, 753–761.
20. Choi, M.; Kim, D.-S.; Kim, Y.H.; Yoon, D.W. Circular cone and its Gauss map. *Colloq. Math.* **2012**, *129*. [CrossRef]
21. Choi, M.; Kim, Y.H.; Yoon, D.W. Classification of ruled surfaces with pointwise 1-type Gauss map. *Taiwan. J. Math.* **2010**, *14*, 1297–1308. [CrossRef]
22. Jung, S.M.; Kim, D.-S.; Kim, Y.H.; Yoon, D.W. Gauss maps of ruled submanifolds and applications *I*. *J. Korean Math. Soc.* **2016**, *53*, 1309–1330. [CrossRef]
23. Kim, D.-S.; Kim, Y.H.; June, S.M.; Yoon, D.W. Gauss maps of ruled submanifolds and applications *II*. *Taiwan. J. Math.* **2016**, *20*, 227–242. [CrossRef]
24. Kim, Y.H.; Yoon, D.W. On the Gauss map of ruled surfaces in Minkowski space. *Rocky Mt. J. Math.* **2005**, *35*, 1555–1581. [CrossRef]
25. Jung, S.M.; Kim, Y.H. Gauss Map and Its Applications on Ruled Submanifolds in Minkowski Space. *Symmetry* **2018**, *10*, 218. [CrossRef]
26. Jung, S.M.; Kim, D.-S.; Kim, Y.H. Minimal ruled submanifolds associated with Gauss map. *Taiwan. J. Math.* **2018**, *22*, 567–605. [CrossRef]
27. Obata, M. The Gauss map of immersions of Riemannian manifolds in space of constant curvature. *J. Differ. Geom.* **1968**, *2*, 217–223. [CrossRef]

28. Chen, B.-Y.; Lue, H.-S. Spherical submanifolds with finite type spherical Gauss map. *J. Korean Math. Soc.* **2007**, *44*, 407–442. [CrossRef]

29. Osserman, R. Minimal surfaces, Gauss maps, total curvature, eigenvalues estimates and stability. In *The Chern Symposium 1979*; Springer: New York, NY, USA, 1980; pp. 199–227.

30. Bekts, B.; Canfes, E.O.; Dursun, U. Pseudo-spherical submanifolds with 1-type pseudo-spherical Gauss map. *Results Math.* **2017**, *71*, 867–887. [CrossRef]

31. Bekts, B.; Canfes, E.O.; Dursun, U. Classification of surfaces in a pseudo-sphere with 2-type pseudo-spherical Gauss map. *Math. Nachrichten* **2017**, *290*. [CrossRef]

32. Bekts, B.; Dursun, U. On spherical submanifolds with finite type spherical Gauss map. *Adv. Geom.* **2016**, *16*, 243–251. [CrossRef]

symmetry

MDPI

Article

Inaudibility of *k*-D'Atri Properties

Teresa Arias-Marco *,† and **José Manuel Fernández-Barroso** †

Departamento de Matemáticas, Universidad de Extremadura, Av. de Elvas s/n, 06006 Badajoz, Spain; ferbar@unex.es
* Correspondence: ariasmarco@unex.es
† These authors contributed equally to this work.

Received: 9 September 2019; Accepted: 17 October 2019; Published: 20 October 2019

Abstract: Working on closed Riemannian manifolds the first author and Schueth gave a list of curvature properties which cannot be determined by the eigenvalue spectrum of the Laplace–Beltrami operator. Following Kac, it is said that such properties are inaudible. Here, we add to that list the dimension of the manifold minus three new properties namely k-D'Atri for $k = 3, ..., \dim M - 1$.

Keywords: Laplace operator; isospectral manifolds; geodesic symmetries; D'Atri space; k-D'Atri space; $\mathfrak{GC}$-space

MSC: 58J50; 58J53; 53C25; 53C20; 22E25; 14J70

1. Introduction

The Inverse Spectral Geometry focus on seeing the unseen [1]. The study of these problems already started in the nineteenth century, inspired by different physical problems. However, the explosion arrived with an affirmative example by M. Kac [2] and a counterexample due to C. Gordon, D. Webb and S. Wolpert in [3], and later with Calderón's problem [4]. The Calderón's problem is also called the Electrical Impedance Tomography Problem in which new advances have been obtained in [5].

These problems can be considered a mix between *Riemannian Geometry*, which studies the geometrical properties on Riemannian manifolds, and *Spectral Geometry*, which focuses on the study of eigenvalue problems. One of the more classical is the closed eigenvalue problem [6].

Let M be a compact and connected manifold without boundary, the solution to this problem is to find all real numbers λ for which there exists a nontrivial solution $f \in C^2(M)$ to the equation

$$\Delta f = \lambda f,$$

where Δ is the Laplace–Beltrami operator acting on functions. The set of real numbers which satisfies the equation is called the eigenvalues of Δ and they form a sequence

$$0 \leq \lambda_1 \leq \lambda_2 \leq \cdots \nearrow \infty.$$

The results presented in this paper contribute to Inverse Spectral Geometry from the classical point of view. That is, it contributes to search which geometrical properties can be determined on closed manifolds by the Laplace spectrum on functions. These properties are said to be audible since Kac's paper's. For example, it is well known that the volume of a closed manifold is spectrally determined. However, in [7] was proved that the following properties among others are inaudible: Weak local symmetry, D'Atri property and the type $\mathcal{A}$ property.

D'Atri spaces were introduced by D'Atri and Nickerson in [8] as a generalization of locally symmetric spaces. The fact that the local geodesic symmetries been volume preserving (up to sign) characterize these spaces. This property became equivalent to the fact that the geodesic symmetries preserve the mean curvature of small geodesic spheres. Two dimensional D'Atri spaces are locally symmetric and then, they have constant sectional curvature. O. Kowalski classified the three dimensional spaces in [9] and he proved that all of them are either locally symmetric or locally isometric to a naturally reductive spaces. In dimension 4, the classification of D'Atri spaces is known only in the locally homogeneous case [10]. Moreover, it is still unknown whether all of them are locally homogeneous (i.e., if the pseudo-group of the local isometries acts locally and transitively on it). Another characterization of D'Atri spaces were proved by D'Atri and Nickerson [8] and improved by Szabó [11] using an infinite series of curvature conditions, namely the Ledger conditions. More precisely, M is a D'Atri space if and only if it satisfies the infinite series of odd Ledger conditions. When a Riemannian manifold satisfies the first odd Ledger condition it is a type $\mathcal{A}$ space or in other words, it has the type $\mathcal{A}$ property. Thus type $\mathcal{A}$ spaces contain D'Atri spaces as a subclass.

On the other hand, k-D'Atri spaces of dimension n, $1 \leq k \leq n - 1$, are a generalization of D'Atri spaces introduced by Kowalski, Prüfer and Vanhecke in [12]. These spaces are those where the geodesic symmetries preserve the k-th elementary symmetric functions of the eigenvalues of the shape operators of all small geodesic spheres. In fact, D'Atri and 1-D'Atri are equivalent conditions and moreover, Druetta proved in [13] that 2-D'Atri is also equivalent to D'Atri condition.

An open question about k-D'Atri spaces is to determine the interrelation between k-D'Atri spaces for different values of k, $k = 2, \ldots, n - 1$, as well as their relation with locally homogeneous spaces. In a different direction, another interesting open question is to determine if the k-D'Atri property can be audible for each value of k. In this paper we solve the last one, given a negative answer for any value of k.

Main Result. *Let M be a Riemannian closed manifold of dimension n, the property of being k-D'Atri for all k, $k = 1, \ldots, n - 1$, cannot be heard.*

Particularly, we obtain directly the following corollary.

Corollary 1. *Under the assumption of the main result, the property of being k-D'Atri for each k, $k = 3, \ldots, n - 1$ is inaudible.*

Two closed Riemannian manifolds are *isospectral* if they have the same eigenvalue spectrum of the Laplace operator acting on functions, counting multiplicities. Thus, a strategy to find an inaudible property is to search two isospectral manifolds which differ from such property. To prove the main result, we will use Szabó manifolds [14]. For the sake of completeness, these manifolds will be presented in detail in Section 3. The needed preliminaries about k-D'Atri spaces will be shown in Section 2. We present the proof of the main results in the last section.

2. About k-D'Atri Properties

Let M be a Riemannian manifold, a point $m \in M$ and a vector $v \in T_m M$, $\|v\| = 1$. We denote by $\gamma_v(r)$ the geodesic in M which starts in m and has initial vector v. Moreover, for each small $r > 0$, we denote by $S_v(r)$ the shape operator of the geodesic sphere

$$G_m(r) = \{\gamma_w(r) = \exp_m(rw) : w \in T_m M, \|w\| = 1\}$$

at $\gamma_v(r)$. For each $m \in M$ the local geodesic symmetry s_m is defined by

$$s_m = \exp_m \circ (-Id) \circ \exp_m^{-1}.$$

An elementary symmetric function σ_k of a symmetric endomorphism A on an n-dimensional real vector space is given by its characteristic polynomial

$$\det(\lambda I - A) = \lambda^n - \sigma_1(A)\lambda^{n-1} + \cdots + (-1)^k \sigma_k(A)\lambda^{n-k} + \cdots + (-1)^n \sigma_n(A)$$

where

$$\sigma_k(A) = \sum_{i_1 < \cdots < i_k} \lambda_{i_1}(A) \cdots \lambda_{i_k}(A)$$

with $1 \leq i_1 < \cdots < i_k \leq n$ and $\{\lambda_1(A), \ldots, \lambda_n(A)\}$ the set of n eigenvalues of A.

Definition 1. *An n-dimensional Riemannian manifold is said to be a k-D'Atri space, $1 \leq k \leq n-1$, if the geodesic symmetries preserve the k-th elementary symmetric functions of the eigenvalues of the shape operator of all small geodesic spheres. That is, for each small $r > 0$ and each unit vector $v \in T_m M$, M is a k-D'Atri space for some $1 \leq k \leq n-1$ if and only if*

$$\sigma_k(S_v(r)) = \sigma_k(S_{-v}(r)).$$

All these spaces are relevant examples of a more general one introduced by Gray in [15].

Definition 2. *We say that a Riemannian manifold M is a type $\mathcal{A}$ space if and only if the Ricci tensor is cyclic parallel, this is*

$$(\nabla_X ric)(X, X) = 0$$

for all $X \in \mathfrak{X}(M)$, where ∇ denotes the Levi-Civita connection.

Proposition 1 ([16]). *If M is a k-D'Atri space then is a type $\mathcal{A}$-space.*

Moreover, when a space has the property of being k-D'Atri for all possible values of k, it has an extra geometrical property.

Proposition 2 ([16]). *M is an n-dimensional k-D'Atri space for all $k = 1, \ldots, n-1$ if and only if for any small real $r > 0$ and any unit vector $v \in T_m M$, the eigenvalues of $S_v(r)$ are preserved by the geodesic symmetries s_m for all $m \in M$, that is*

$$ds_m|_{\gamma_v(r)} \circ S_v(r) = S_{-v}(r) \circ ds_m|_{\gamma_v(r)}.$$

This property was introduced by J. Berndt, F. Prüfen and L. Vanhecke in [17] and namely $\mathfrak{GC}$-property.

3. The Riemannian Manifolds $N^{(a,b)}$

Now we are going to expose $N^{(a,b)}$, the Szabó manifolds [14], as a special class of the manifolds $N(j)$ introduced in [18]. To construct $N(j)$ we need:

1. A two step nilpotent Lie algebra $\mathfrak{g}(j) = \mathfrak{v} \oplus \mathfrak{z}$ with an inner product for which $\mathfrak{v}$ and $\mathfrak{z}$ are orthogonal, where $\mathfrak{z}$ is central, $j : \mathfrak{z} \to \mathfrak{so}(\mathfrak{v})$ is a linear map and the Lie bracket $[\cdot, \cdot] : \mathfrak{v} \times \mathfrak{v} \to \mathfrak{z}$ is given by the equation

$$\langle [X, Y], Z \rangle = \langle j_Z X, Y \rangle, \quad X, Y \in \mathfrak{v}, \; Z \in \mathfrak{z}.$$

The Lie algebra $\mathfrak{g}(j)$ has an associated two-step simply connected nilpotent Lie group $\tilde{G}(j)$ defined by the exponential map, $\exp : \mathfrak{v} \oplus \mathfrak{z} \to \tilde{G}(j)$ by $\exp(X, Z) = (X + Z)$. Its Lie group multiplication is given by the Campbell-Baker-Hausdorff formula as follows

$$\exp(X, Z) \cdot \exp(Y, W) = \exp\left(X + Y, Z + W + \frac{1}{2}[X, Y]\right).$$

Please note that the inner product on the Lie algebra $\mathfrak{g}(j)$ defines a left-invariant metric on the Lie group $\tilde{G}(j)$, that is a metric for which the left translations by group elements are isometries.

2. We consider the submanifold of $\tilde{G}(j)$ without boundary

$$\tilde{N}(j) = \left\{ \exp(X, \tilde{Z}) \in \tilde{G}(j) : X \in S^{\dim \mathfrak{v} - 1} \text{ and } \tilde{Z} \in \mathfrak{z} \right\} \cong S^{\dim \mathfrak{v} - 1} \times \mathfrak{z}.$$

3. Now, to obtain a closed manifold, we take a lattice $\mathcal{L}$ of full rank in $\mathfrak{z}$ and we consider $G(j) = \tilde{G}(j) / \exp(\mathcal{L})$.

4. Finally, we obtain the closed submanifold

$$N(j) = \left\{ \exp(X, Z) \in G(j) : X \in S^{\dim \mathfrak{v} - 1} \text{ and } Z \in \mathfrak{z} / \mathcal{L} \right\} \cong S^{\dim \mathfrak{v} - 1} \times T^{\dim \mathfrak{z}}.$$

This construction gives us the following diagram

$$
\begin{array}{ccccc}
\mathfrak{g}(j) & \xrightarrow{\exp} & \tilde{G}(j) & \rightsquigarrow & G(j) & = & \tilde{G}(j) / \exp(\mathcal{L}) \\
& & \cup & & \cup & & \\
& & \tilde{N}(j) & \rightsquigarrow & N(j) & = & \tilde{N}(j) / \exp(\mathcal{L}),
\end{array}
$$

where $\rightsquigarrow$ denotes a Riemannian covering. Please note that the tangent space of $\tilde{N}(j)$ at some $p = \exp(x, z) \in \tilde{N}(j)$ with $x \in \mathfrak{v}$, $\|x\| = 1$, $z \in \mathfrak{z}$, is given by

$$T_p \tilde{N}(j) = L_{p*} \left\{ (X, Z) : X \in \mathfrak{v}, \ X \perp x, \ Z \in \mathfrak{z} \right\}.$$

Moreover, $N(j)$ has constant scalar curvature (see [18]).

To get the Szabó manifolds we need to consider the next particular map j. Let $\mathbb{H} = \operatorname{span} \{1, \mathbf{i}, \mathbf{j}, \mathbf{k}\}$ be the algebra of quaternions with the usual multiplication. For $a, b \in \mathbb{N}_0$ with $a + b > 0$, we define $\mathfrak{v}$ as the direct orthogonal sum of $a + b$ copies of $\mathbb{H}$. Let $\mathfrak{z} = \operatorname{span} \{\mathbf{i}, \mathbf{j}, \mathbf{k}\}$, $\mathcal{L} = \operatorname{span}_{\mathbb{Z}} \{\mathbf{i}, \mathbf{j}, \mathbf{k}\}$ and the linear map $j^{(a,b)} : \mathfrak{z} \to \mathfrak{so}(\mathfrak{v})$ defined by

$$j_Z^{(a,b)}(X_1, \ldots, X_a, X_{a+1}, \ldots, X_{a+b}) := (X_1 Z, \ldots, X_a Z, Z X_{a+1}, \ldots, Z X_{a+b}).$$

Finally, we denote $N^{(a,b)} = N(j^{(a,b)})$, respectively $\tilde{N}^{(a,b)} = \tilde{N}(j^{(a,b)})$.

Now, we are interested in finding pairs of isospectral manifolds inside the class of $N^{(a,b)}$. The next result is essential.

Proposition 3 ([18]). *If two linear maps $j, j' : \mathfrak{z} \to \mathfrak{so}(\mathfrak{v})$ have the same eigenvalues counting multiplicities in $\mathbb{C}$, then the closed Riemannian manifolds $N(j)$ and $N(j')$ are isospectral for the Laplace operator on functions.*

Please note that $j^{(a,b)}$ is of Heissenberg type, hence for $j_Z^{(a,b)^2} = -\|Z\|^2 Id_{\mathfrak{v}}$. Thus, their eigenvalues are $\pm i \|Z\|$, each with multiplicity $\dim \mathfrak{v} / 2$.

Corollary 2. *Two submanifolds $N^{(a,b)}$ and $N^{(a',b')}$ are isospectral if and only if $a + b = \dim \mathfrak{v} / 4 = a' + b'$.*

Moreover, the pair of isospectral manifolds $N^{(a+b,0)}$ and $N^{(a,b)}$, $b \geq 0$, are an optimal pair to study the audibility of k-D'Atri spaces because they also have the following property that proves the inaudibility of the local homogeneous property.

Proposition 4 ([14]). *$N^{(a+b,0)}$ are locally homogeneous while $N^{(a,b)}$, $b \geq 0$ are not.*

4. Proof of Main Results

Weakly symmetric spaces were introduced by Selberg in [19]. Szabó in [11] introduced a new definition which was called *ray symmetric spaces*. Then, Berndt and Vanhecke proved in [20] that these two definitions are equivalent. A Riemannian manifold is called weakly symmetric (in the sense of Szabó) if for each $m \in M$ and each nontrivial geodesic γ starting in m, there exists an isometry f of M which fixes m and reverses γ, that is

$$df_m(\dot{\gamma}(0)) = -\dot{\gamma}(0).$$

Related with this kind of spaces, it is well known the following result.

Proposition 5 ([17]). *Every weakly symmetric space is a $\mathfrak{GC}$-space.*

A Riemannian manifold is *weakly-locally symmetric* (see [7]) if for every $m \in M$ there exists $\varepsilon > 0$ such that for any unit speed geodesic γ in M with $\gamma(0) = m$ there exists an isometry of the distance ball $B_\varepsilon(m)$ which fixes m and reverses $\gamma|_{(-\varepsilon,\varepsilon)}$. With this definition we have the following consequences.

Lemma 1 ([7]). *Let M be a complete, simply connected, weakly-locally symmetric Riemannian manifold. Then M is weakly symmetric. In particular, the universal Riemannian covering of any complete, weakly-locally symmetric Riemannian manifold is weakly symmetric.*

Now, let us focus on checking the property of being k-D'Atri on Szabó manifolds.

As is shown in [7], the manifolds $N^{(a+b,0)}$ are weakly locally symmetric for any $a, b \in \mathbb{N}_0, a + b > 0$.

Therefore, $\tilde{N}^{(a+b,0)}$ are weakly symmetric by the previous Lemma and they are $\mathfrak{GC}$-spaces by Proposition 5. Finally, using Proposition 2, $\tilde{N}^{(a+b,0)}$ are k-D'Atri spaces for all k, $k = 1, \ldots, n-1$. Now, $N^{(a+b,0)}$ inherits this property because it is a local property and these two Riemannian manifolds are locally isometric. Thus, $N^{(a+b,0)}$ are k-D'Atri for all k.

On the other hand, it is known that $\tilde{N}^{(a,b)}$ are not type $\mathcal{A}$-spaces by [7].

Therefore, using Proposition 2, $\tilde{N}^{(a,b)}$ are not k-D'Atri for any k. Moreover, $N^{(a,b)}$ neither satisfy the property of being k-D'Atri for any k because $\tilde{N}^{(a,b)}$ is its universal Riemannian covering and the property is local.

Then, we have two isospectral manifolds, $N^{(a+b,0)}$ and $N^{(a,b)}$, one of them is k-D'Atri for all $k = 1, \ldots, n-1$ and the other is not k-D'Atri for any possible value of k.

The proof of the Corollary 1 is now immediate from the fact that if $N^{(a+b,0)}$ is k-D'Atri for all k, it is in particular for each k.

5. Conclusions and Applications

Inverse spectral geometry is based on determining the shape and properties of unknown objects using the least amount of information, for example, with only the spectrum of a determined operator.

Following Kac [2], it is said that the properties which can be recovered by the spectrum of the Laplace–Beltrami operator are audible.

From a more applied point of view, inverse problems have to do with moving from effect to cause. Therefore, the treatment of these problems is both mathematical and computational. Given a certain measurement data from an unknown object of interest, the point is to design a computational algorithm that takes the data as input and produces, for example, an image of the unknown object. There are some operators whose applications are already a reality, such as the Dirichlet-to-Neumann operator for which already exists an experimental team developing its applications in Electrical Impedance Tomography with promising advances in the detection of breast cancer.

The main result proved in this paper provides us the fact that one cannot determine by the eigenvalues of the Laplace–Beltrami operator if a Riemannian closed manifold is k-D'Atri or not, for each possible value of k.

Therefore, a computational algorithm cannot be designed to determine these properties. This will avoid the costs of creating an applied study in relation to the property of being k-D'Atri.

Author Contributions: All authors contributed equally to this research and in writing the paper.

Funding: The authors are partially supported by Junta de Extremadura and Fondo Europeo de Desarrollo Regional (GR18001 and IB18032). The first author is also supported by Dirección General de Investigación Española and Fondo Europeo de Desarrollo Regional (MTM2016-77093-P).

Acknowledgments: The first author is delighted to thanks conversations with her Master student Paloma Megías Mesa.

Conflicts of Interest: The authors declare no conflict of interest. The funders had no role in the design of the study; in the collection, analyses, or interpretation of data; in the writing of the manuscript, or in the decision to publish the results.

References

1. Uhlmann, G. Inverse problems: Seeing the unseen. *Bull. Math. Sci.* **2014**, *4*, 209–279. [CrossRef]
2. Kac, M. Can one hear the shape of a drum? *Am. Math. Mon.* **1966**, *73*, 1–23. [CrossRef]
3. Gordon, C.; Webb, D.; Wolpert, S. One cannot hear the shape of a drum. *Bull. Am. Math. Soc.* **1992**, *27*, 134–138. [CrossRef]
4. Calderón, A.P. On an inverse boundary value problem. In *Seminar on Numerical Analysis and Its Applications to Continuum Physics (Rio de Janeiro)*; Polytechnic University of Turin: Turin, Italy, 1980; pp. 65–73.
5. Arias-Marco, T.; Dryden, E.B.; Gordon, C.S.; Hassannezhad, A.; Ray, A.; Stanhope, E. Spectral geometry of the Steklov problem on Orbifolds. *Int. Math. Res. Not. IMRN* **2019**, *1*, 90–139. [CrossRef]
6. Chavel, I. *Eigenvalues in Riemannian Geometry*; Academic Press: Cambridge, MA, USA, 1984.
7. Arias-Marco, T.; Schueth, D. On inaudible properties of closed Riemannian manifolds. *Ann. Glob. Anal. Geom.* **2010**, *4*, 339–349. [CrossRef]
8. D'Atri, J.E.; Nickerson, H.K. Geodesic symmetries in spaces with special curvature tensors. *J. Differ. Geom.* **1974**, *9*, 251–262. [CrossRef]
9. Kowalski, O. Spaces with volume-preserving symmetries and related classes of Riemannian manifolds. *Rend. Sem. Mat. Univ. Politec. Torino Fascicolo Speciale* **1984**, 131–158. Available online: https://ci.nii.ac.jp/naid/10003478200/ (accessed on 9 September 2019).
10. Arias-Marco, T.; Kowalski, O. Classification of 4-dimensional homogeneous D'Atri spaces. *Czechoslovak Math.* **2008**, *58*, 203–239. [CrossRef]
11. Szabó, Z.I. Spectral theory for operator families on Riemannian manifolds. *Proc. Symp. Pure Math.* **1993**, *3*, 615–665.
12. Kowalski, O.; Prüfer, F.; Vanhecke, L. D'Atri spaces. *Prog. Nonlinear Differ. Equ. Appl.* **1996**, *20*, 241–284.
13. Druetta, M.J. Geometry of D'Atri spaces of type k. *Ann. Glob. Anal. Geom.* **2010**, *38*, 201–219. [CrossRef]
14. Szabó, Z.I. Locally non-symmetric yet super isospectral spaces. *Geom. Funct. Anal.* **1999**, *9*, 185–214. [CrossRef]
15. Gray, A. Einstein-like manifolds which are not Einstein. *Geom. Dedic.* **1978**, *7*, 259–280. [CrossRef]
16. Arias-Marco, T.; Druetta, M.J. D'Atri spaces of type k and related classes of geometries concerning Jacobi Operators. *J. Geom. Anal.* **2014**, *24*, 721–739. [CrossRef]
17. Berndt, J.; Prüfer, F. Vanhecke, L. Symmetric-like Riemannian manifolds and geodesic symmetries. *Proc. R. Soc. Edinb. Sect. A* **1995**, *125*, 265–282. [CrossRef]
18. Gordon, C.S.; Gornet, R.; Schueth, D.; Webb, D.; Wilson, E.N. Isospectral deformations of closed Riemannian manifolds with different scalar curvature. *Ann. Inst. Fourier* **1998**, *48*, 593–607. [CrossRef]
19. Selberg, A. Harmonic analysis and discontinuous groups in weakly symmetric Riemannian spaces with applications to Dirichlet series. *J. Indian Math. Soc.* **1956**, *20*, 47–87.
20. Berndt, J.; Vanhecke, L. Geometry of weakly symmetric spaces. *J. Math. Soc. Jpn.* **1996**, *48*, 745–760 [CrossRef]

MDPI

St. Alban-Anlage 66

4052 Basel

Switzerland

Tel. +41 61 683 77 34

Fax +41 61 302 89 18

www.mdpi.com

Symmetry Editorial Office

E-mail: symmetry@mdpi.com

www.mdpi.com/journal/symmetry

www.ingramcontent.com/pod-product-compliance
Lightning Source LLC
Chambersburg PA
CBHW051912210326
41597CB00033B/6119